W9-BIN-784

DO-IT-YOURSELF YEARBOOK 2000

Hearst Communications, Inc.

This book is published with the consent and cooperation of POPULAR MECHANICS Magazine.

For POPULAR MECHANICS:
Editor-in-Chief: Joe Oldham
Creative Director: Bryan Canniff
Managing Editor: Sarah Deem
Home Improvement Editors: Steven Willson, Thomas Klenck, Roy Berendsohn
Technology Editor: Tobey Grumet
Science Editor: Jim Wilson
Automotive Editors: Don Chaikin, Jim Dunne, Mike Allen, Scott Oldham
Photo Editor: Nancy Coggins

http://popularmechanics.com

POPULAR MECHANICS 2000 YEARBOOK
Editor: Paul Currie
Book Design and Production: Bill Nelson
Editorial Assistance: Stacy Sheagley
Cover Photo: Rosario Capotosto
Cover Design: Bill Nelson
Hearst Direct Marketing Staff: Mindy Francus

ISBN 0-688-16137-5
ISSN 1097-2781
GST Hearst Registration No. R10521891

10 9 8 7 6 5 4 3 2 1
Printed in the United States of America

Although every effort has been made to ensure the accuracy and completeness of the information in this book, Hearst Direct Books and Popular Mechanics Magazine make no guarantees, stated or implied, nor will they be liable in the event of misinterpretation or human error made by the reader or for any typographical errors that may appear. Plans for projects illustrated in this book may not meet all local zoning and building code requirements for construction. Before beginning any major project, consult with local authorities or see a structural architect. WORK SAFELY WITH HAND TOOLS. WEAR EYE PROTECTION. READ MANUFACTURER'S INSTRUCTIONS AND WARNINGS FOR ALL PRODUCTS.

Introduction

The year 2000 brings many exciting changes and opportunities—for your home as well as your life. Another year brings transitions in your or your family's lifestyle, plus more wear and tear on your house. This new volume in the Popular Mechanics Yearbook series provides projects that will help you meet these changing needs.

It offers exciting furniture projects to enhance the comfort, style and function of your home. Plus there are beautiful home improvement projects—a traditional picket fence, a stunning cobblestone patio and barbecue combination and an amazing porch addition—that will enrich your home's looks as well as your entertainment possibilities. Also, keeping up with repair needs are window and shingle replacement projects, and simple but common repairs on two significant home appliances, the refrigerator and washing machine, to keep you functional through the year 2000 and beyond.

Step-by-step instructions guide you through all of the projects. Each of the building projects is accompanied by a technical illustration that shows the project parts in an exploded diagram, making it simple to see how the piece fits together. A list of materials and part sizes is included where appropriate.

Additionally there is a large section providing information on setting up a workshop with the tools and materials you need to begin building projects. You see how to build a collection of basic tools that will allow you to build just about anything. Plus you learn about choosing wood stock, glues and abrasives for your projects, as well as basic techniques for joining project parts together. Even if you're an accomplished woodworker you'll find this helpful reference material.

May the year 2000 be a great one for you, and we hope this book will increase your enjoyment of it.

Table of Contents

Getting Started

If you've always wanted to build fine furniture, why not start now? It's not that difficult to test the waters and it doesn't have to be that expensive. You just have to be willing—if you want to increase your chances of success—to begin at the beginning. Like any other new pursuit, the learning curve tends to be pretty steep at the start. But this curve can level off pretty quickly, especially if you have a firm foundation in the basics.

At first, the whole process may seem a bit mysterious. How does the idea of a table, for example, get from your head to your living room in just a few weeks' time? Well, it gets there following a pretty logical sequence. First you need a plan. You can come across such drawings in books like this one (see "Console Table," page 68) or from magazines or plan services—or simply by taking careful measurements from an existing piece.

Then you have to buy enough tools to ensure success. You can't prevail with just a screwdriver and a chisel. But you certainly don't need $20,000 worth of shop equipment either. "Beginner's Toolbox," on page 6, shows you everything you need to get started. As you'll see, this selection includes only a few common portable power tools. The rest are hand tools. Not only does this keep your initial investment low, it also makes you focus on proper hand tool use, which many professionals think is the basis of true craftsmanship.

With plan and tools on hand, your next step is to buy the wood and supplies necessary to complete the job. To this end we've included stories on proper stock selection and preparation as well as information on choosing and using woodworking glues and abrasives. We also touch on basic sharpening techniques, because dull tools can frustrate you more easily than anything else in the whole process.

Of course the core discipline is, and will always be, joinery. In our story on page 37 we cover the basics, knowing full well that we're only scratching the surface. But if you can master what we show, more exotic joinery will be much easier to pursue in the future. The same can be said for our finishing story, which begins on page 41. The number of ways you can protect your finished piece are legion. But you don't have to learn them all at once.

Finally, we offer you a detailed story on building a solid cherry console table. The complete plan mentioned earlier is included, as well as a series of step-by-step photos that clearly explain the whole process. Everything you need is in this story and the ones that precede it. This piece may be simple but that doesn't mean it's not well-designed. Its straightforward, elegant lines and

its timeless joinery yield a piece that's well worth the effort required to make it.

So, why not give it a whirl? Furniture making is fun, and doesn't cost much compared to many other pursuits. And the fruits of your labor will be there for you, and everyone else, to see.

Beginner's Toolbox

Just about any furniture maker will tell you that it's hard to have too many tools. This is probably true because these people have a deep appreciation for just how useful tools can be. The right one for the job can make any task easier, quicker and very often safer. Over the years a successful artisan can accumulate a truly astounding quantity of equipment and supplies. Each is important, or at least was at one time, and therefore difficult to leave behind. If you're a professional this makes perfect sense. But for the beginner such conspicuous consumption can be a real mistake. Why devote so many resources up front to something you're trying for the first time?

A better idea is to buy a basic assortment of woodworking tools like the one shown here. With these tools you can build just about any straightforward furniture piece and leave the extra room under your credit limit for other things. This is not to say that tool manufacturers are giving their wares away. For the full complement shown, expect to pay around $800, depending on the quality of the individual items. This is a lot of money. But it's also a lot of capability.

When shopping, it's a good idea to buy the best tool you can afford in any category. Do keep in mind that price may not always be the best indicator of quality, but it usually is. Cheap tools often have a much shorter life, and are typically less accurate and frequently

more dangerous to use than their premium brothers and sisters. Also remember that although some of these tools are fairly specialized, most of them can be used for general repair chores around the house. So even if you leave the world of furniture making behind, your investment in these tools will reap benefits for years to come.

Of course, this tool selection is abbreviated on many fronts. For example, we haven't presented some common tools that most people already have around the house. These include a 16-ounce claw hammer, a heavy-duty, 25-ft. grounded extension cord, an assortment of flat and Phillips screwdrivers, a tape measure, safety glasses and hearing protectors. We also assume that most people have some kind of workbench or worktable that can be used for support. If you don't, a convenient and economical option is a folding Workmate. These units cost about $90. They not only provide sturdy support but also boast movable clamping jaws that can hold just about anything securely in place.

So, on to the tools themselves. Following is a brief description of each, along with the approximate cost of high-quality models.

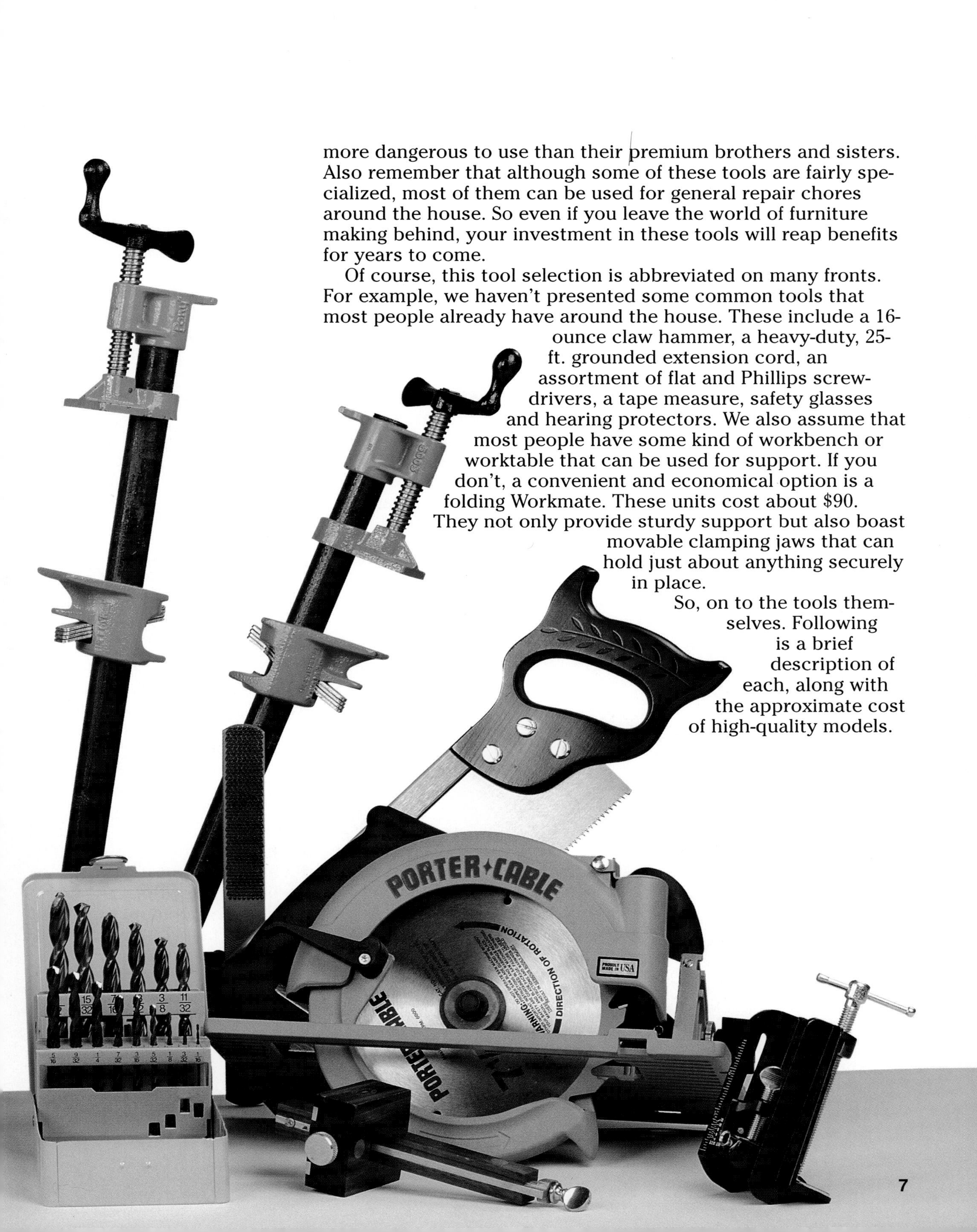

Circular Saw

A circular saw is a versatile tool for both rough and finish cutting. The saw consists of a horizontally mounted motor that drives a 7¼-in.-dia. blade. The depth of the cut, as well as the bevel angle, are adjustable. Many saws come equipped with an accessory rip guide, but if yours doesn't you should buy one. Also, outfit your saw with a carbide-tipped, thin-kerf blade. This will cut at least 10 times as long as a steel blade before requiring sharpening. And it reduces the load on the saw motor and wastes less stock to sawdust. This is the first tool to pick up for both crosscutting (perpendicular to the grain) and ripping (cutting parallel to the grain) solid stock as well as for sizing manufactured panels. The spinning saw blade enters the bottom side of the workpiece, which can result in chip-out on the top surface. Plan your cuts so that the good side of your material faces down during cutting. **(Approximate price $130.)**

Drill and Bits

Most woodworking projects require you to bore holes of some sort and a drill is the only tool for this job. Handheld electric drills are commonly available in three sizes (¼ in., ⅜ in. and ½ in.) that represent the maximum bit diameter the drill chuck will accept. While the ½-in. drill is the most versatile, it also tends to be the heaviest and most awkward to use. A good compromise for the beginning woodworker is to purchase a ⅜-in. VSR (variable speed reversible) drill. The newer cordless models are rated at 14.4 volts and offer substantial power and convenience—but at a substantially higher price than corded models. An assortment of high-speed steel twist drill bits from 1⁄16 in. to ½ in. dia. will cover most needs.

The addition of a set of countersinks for recessing screwheads will allow you to make even better use of the drill. **(Approximate prices: corded drill, $100; cordless drill, $180; drill bits, $30.)**

Block and Bench Planes

Quality planes are some of the most versatile woodworking tools, and often the most satisfying to use. The sound and feel of a sharp plane slicing through wood is, to many, the essence of woodworking. A plane consists of a sharpened steel

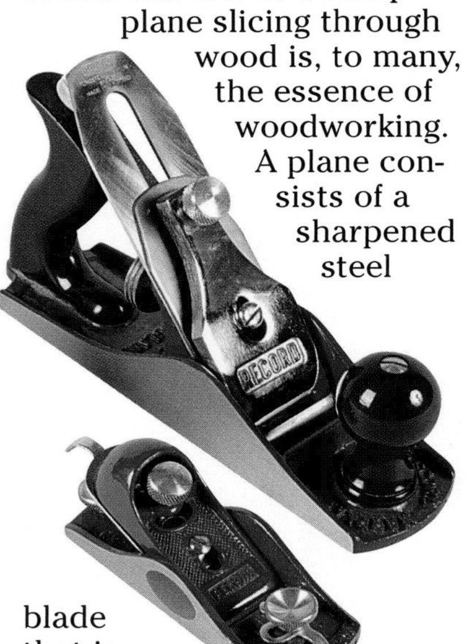

blade that is held at a fixed angle in a steel or wood body. The blade, or iron, is adjustable to regulate the depth of cut. There are planes designed for general work and planes designed for one particular use. For a beginner, a block plane and a No. 4 bench plane (approximately 9½ in. long) will cover most situations. The block plane is designed to trim end grain but it can also be used any time a bench plane would be too unwieldy. The bench plane is used to square and straighten lumber edges for gluing or to smooth the surface of a board or glued-up panel. Most planing should be done with the tool moving parallel to the grain of the wood. Occasionally you will notice that the tool seems to tear out the wood grain. When this happens, simply work from the opposite direction. (Approximate prices: block plane, $50; bench plane, $75.)

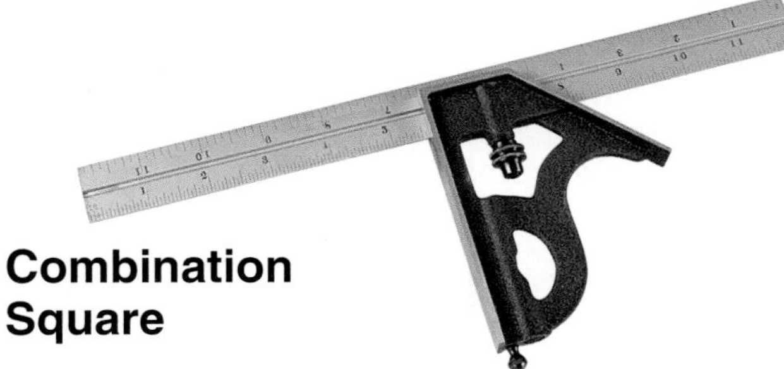

Combination Square

The combination square consists of a cast body that slides along a graduated metal blade. The body can be fixed by a tightening screw at any position along the blade. It provides an accurate standard for either a 90° or 45° mark. Many combination squares incorporate a small level in the tool as well. Use the square to mark lines for cutting and to check that finished cuts are square. The blade can also be removed from the body and used as an accurate layout tool. Since the reliability of this square is so critical to quality work, it's worthwhile to purchase a precision model, like the Starrett shown. Squares are commonly available in 4-in., 6-in. and 12-in. sizes. (Approximate price $50.)

Backsaw

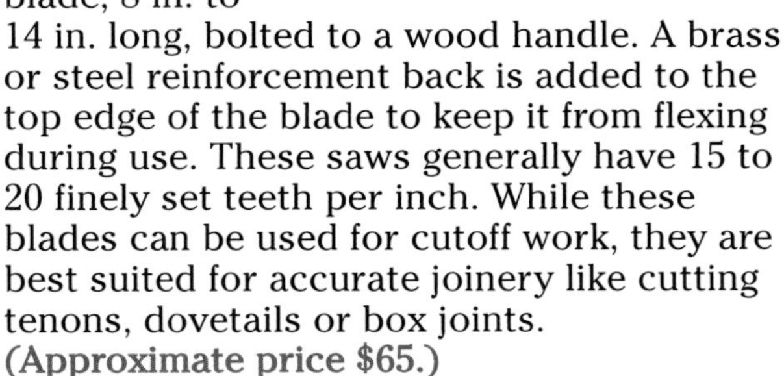

A backsaw consists of a straight blade, 8 in. to 14 in. long, bolted to a wood handle. A brass or steel reinforcement back is added to the top edge of the blade to keep it from flexing during use. These saws generally have 15 to 20 finely set teeth per inch. While these blades can be used for cutoff work, they are best suited for accurate joinery like cutting tenons, dovetails or box joints. (Approximate price $65.)

Chisels

After the knife, a chisel is the most basic of cutting tools. A steel blade of specified width and length, usually from 3 in. to 8 in., is mounted in a wood or plastic handle. A bevel is ground on the end of the blade at an angle varying from 15° to 35°, depending on the intended use. A chisel can be used either to cut with the grain (pare) or to cut across the grain (chop). Plastic-handled chisels with a steel striking plate can be driven either by hand or by striking them with a hammer or mallet. Wood-handled chisels should never be struck with a hammer because this would destroy the handle. Chisels are made in many styles, each for a specific use, but for our purposes an assortment of four butt chisels in widths ranging from ¼ in. to 1 in. is a good place to start. (Approximate price $50.)

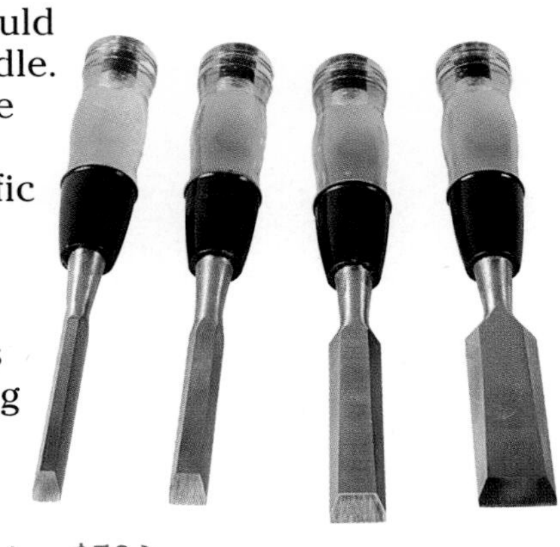

Sharpening Stone and Guide

Keeping a sharp edge on chisels and plane irons is absolutely necessary for the successful and safe use of these tools. A sharpening stone and honing guide are required for this task. While many sharpening systems are available, one of the best is a combination waterstone. The Norton Co. manufactures stones with 220/1000 grit and 1000/4000 grit combinations. For a beginner's all-around use the 220/1000 grit is the best choice. The stone should be soaked in water for about 15 minutes before use and kept wet during sharpening. The water keeps the metal particles from becoming embedded in the stone and glazing its surface. A honing guide is a jig that holds a chisel or plane iron at a constant angle against the stone. To sharpen a tool, move it back and forth on the stone to form the cutting edge. (Approximate prices: stone, $35; guide, $25.)

Doweling Jig

Dowels are one of the simplest and best means for assembling a joint or aligning two

adjacent surfaces. A doweling jig provides a guide for accurately boring the required holes. These jigs come with a variety of bushings, usually ranging in diameter from 1/8 in. to 1/2 in., and a clamping arrangement that holds the desired bushing in position over the workpiece. A drill is used to drive the appropriately sized bit through the bushing to form a hole. A stop is normally attached to the bit to limit the depth of the hole. **(Approximate price $40.)**

Marking Gauge

A combination mortise and marking gauge is extremely useful for joint layout. One side of the gauge has two adjustable pins for scribing the parallel lines needed for mortise or tenon cuts. The opposite side of the gauge has a single pin for general marking, either parallel to or across the grain. **(Approximate price $40.)**

Orbital Sander

Sanding is an essential part of the finishing process for almost any woodworking project. While sanding is probably the least popular aspect of woodworking, it does not need to be tedious or unpleasant. A 1/4 sheet orbital palm sander makes this task relatively painless. An assortment of sandpaper in grits of 100, 120, 150, 180 and 220 will cover most sanding needs. Aluminum oxide paper will provide the longest use and prove to be the most economical choice. It's also wise to have a selection of three grades of steel wool, No. 00, No. 000 and No. 0000, on hand for use during most finishing procedures. **(Approximate price $60.)**

Combination Rasp/File

If you are interested in doing any carving, or including shaped work of any kind in your projects, a combination rasp/file is a valuable addition to your collection. This tool combines a fine and medium rasp with a fine and medium file—to create four separate cutting surfaces. Rasps have small, individual cutting teeth that cut aggressively into wood. The files have a series of parallel and sometimes diagonally opposed ridges to smooth the wood. Used in combination, these tools can quickly shape either simple or complex forms. After using a rasp and file, the next step is to sand the work to a smooth finish. **(Approximate price $15.)**

Clamps

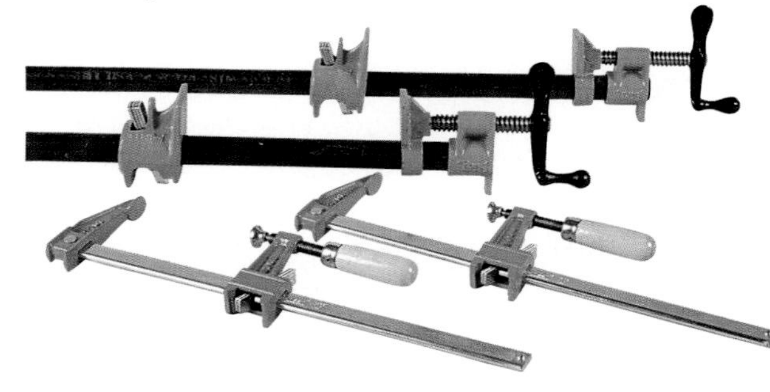

Unless your project parts are fastened with nails or screws, clamps will be required to pull joints tight and hold them while the glue sets. Clamps can be quite expensive—and keep in mind that most experienced woodworkers claim that they never have enough of them. Small quick clamps are good for light assembly and for holding a jig or straightedge in place. Larger quick clamps are for general assembly and laminating solid stock. Pipe clamps, which are made from clamp fixtures and standard black pipe, are necessary for panel, face frame and general furniture assembly. A handy feature of the pipe clamp system is that the fixtures can be removed from one pipe and threaded onto another of a different length, depending on your needs. Two 12-in. quick clamps and four pipe clamps are a good place to begin. (Approximate prices: quick clamps, $15 each; pipe clamp fixtures, $20 each; 5-ft. black pipe, $10.)

Straightedge

Most woodworking procedures begin by establishing one straight edge on each piece of stock. A combination clamp/straightedge tool can serve as a reference point for establishing that edge. If you are interested in doing quality work, this tool will become extremely important in your shop. It is a good idea to purchase the longest and most accurate model you can afford. A 50-in. tool, like the one shown here, is a great place to start. (Approximate price $40.)

The preceding tools will accommodate most simple projects. But if you want to tackle a piece that includes a lot of curved cuts, then you should add an electric jigsaw to your collection right away.

Text and Photos by Neal Barrett

Source List

While most of these tools are readily available at neighborhood hardware stores or home centers, you can also get them from mail-order suppliers that specialize in woodworking tools. Here are the names and addresses of some good ones.

Constantine's, 2050 Eastchester Rd., Bronx, NY 10461; 800-223-8087

Craftsman, Sears Power and Hand Tools, 3333 Beverly Rd., Hoffman Estates, IL 60179; 800-377-7414

Garrett Wade, 161 Avenue of the Americas, New York, NY 10013; 800-221-2942

Lee Valley, 12 E. River St., Ogdensburg, NY 13669; 800-871-8158

Rockler Woodworking And Hardware, 4365 Willow Dr., Medina, MN 55340; 800-279-4441

Woodcraft, 210 Wood County Industrial Park, P.O. Box 1686, Parkersburg, WV 26102; 800-225-1153

Woodworker's Supply, 1108 N. Glenn Rd., Casper, WY 82601; 800-645-9292

Sharpening

There's no question that sharp tools are a requirement for quality work. And the tools that are most apt to dull in frequent use are chisels and planes. These tools are also the easiest ones for beginners to sharpen. Of course, saw blades of all descriptions will also dull over time. But they are more difficult to sharpen and require specialized tools and techniques not normally at the disposal of the beginner. Renewing these blades is better left to a professional sharpening service.

The tools required for sharpening are pretty basic. First, and by far the most important, is the stone. For best results, we recommend man-made waterstones instead of the more familiar oil stones. Waterstones are soft and therefore cut faster. And they don't clog as easily. For both reasons, they're the logical choice for the inexperienced. These stones are available either in single grits or as combination stones of two grits, one on each side. A good stone to start with has a combination of 220- and 1000-grit surfaces.

As the name implies, water is used as a lubricant on these stones. The stone must be soaked in water for about 15 minutes before you use it. And it must be kept wet while you're sharpening. The water prevents the metal filings from becoming embedded in the stone and glazing the surface. When you are done working with a stone, you should rinse it off and dry it before putting it away. If you store the stone in water, it will decompose over time.

Sharpening chisel blades and plane irons requires that you hold the tool at a fixed angle while moving it across the stone. The bevel angle for chisels and plane irons can vary, but for general work a 25° angle is a good compromise.

It is possible to work on a stone holding the tool by hand. Indeed, some experienced furniture makers take a great deal of pride in this skill. But for the beginner, it makes more sense to use a simple honing guide to keep the tool at the proper angle. The guide we show is manufactured by Veritas Tools and is available from Lee Valley and other mail-order tool companies. It comes with a companion angle guide that makes setting the bevel a breeze. One of the best features of this guide is that it rides directly on the stone, allowing you to move from stone to stone without readjusting the guide. It also has a fine adjustment for honing a secondary, micro-bevel on the tool edge that's 1° or 2° steeper than the primary bevel. Because you're taking off less material on the micro-bevel, it's easier and faster to renew the cutting edge when it gets dull.

Be aware that sharpening can be a messy procedure. The process creates a slurry that must be wiped off the tool often to let you check your progress. It's a good idea to place a piece of scrap plywood or rubber under the stone to contain the mess.

Flattening the Back

Sharpening a chisel or plane iron begins with flattening the back of the blade. This side of the tool is rarely flat when it comes from the factory. But flatness is crucial if you want to achieve a razor-sharp edge.

1 Place the blade flat on the 220-grit side and repeatedly rub the blade across the stone *(see Figure A)*. Try to cover the entire surface with your strokes, to maintain even wear on the stone. Continue until you see even scratches across the entire blade surface.

2 Repeat the process on the 1000-grit side.

Primary Bevel

1 Place the blade loosely in the honing guide and adjust its position to create the proper primary bevel *(see Figure B)*. Remember that 25° is a good all-purpose angle.

2 Check that the blade is square to the edges of the honing guide, using a combination square *(see Figure C)*. If the blade isn't square, the sharpened edge won't be either.

3 Carefully examine the edge of the blade. If it's extremely worn, start sharpening on the 220-grit side *(see Figure D)*. If the blade just needs a little touching up, you can use the 1000-grit stone.

4 With your fingertips, place light pressure near the blade's edge and stroke the blade back and forth over the stone. Once again, be sure to use the whole surface of the stone so that the wear is not confined to one area. And be sure to keep the stone soaked with water while you work.

5 Continue until you see a small burr on the back of the blade. This burr is sometimes called a wire edge, and it indicates that you are ready to move onto a finer grit. If you started with the coarse stone, move to the 1000-grit stone and follow the same procedure as before *(see Figure E)*.

FIGURE A: Begin sharpening a chisel by flattening the back of the blade on the coarse side of the waterstone.

FIGURE B: Place the chisel in the guide and set it on the angle jig. Once the angle is adjusted, tighten the blade in the guide.

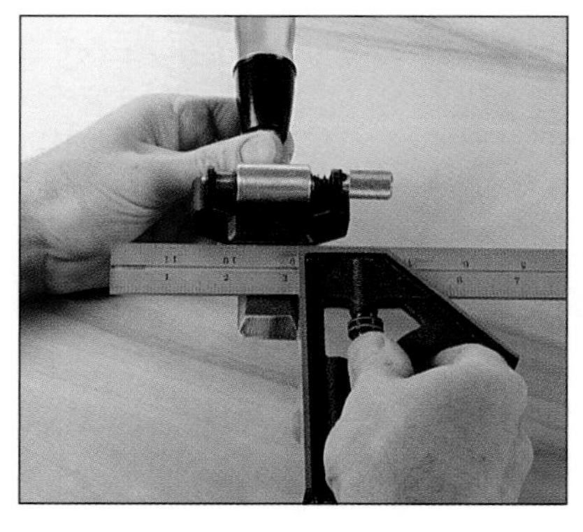

FIGURE C: Check that the chisel is perfectly square to the edges of the honing guide, using a combination square.

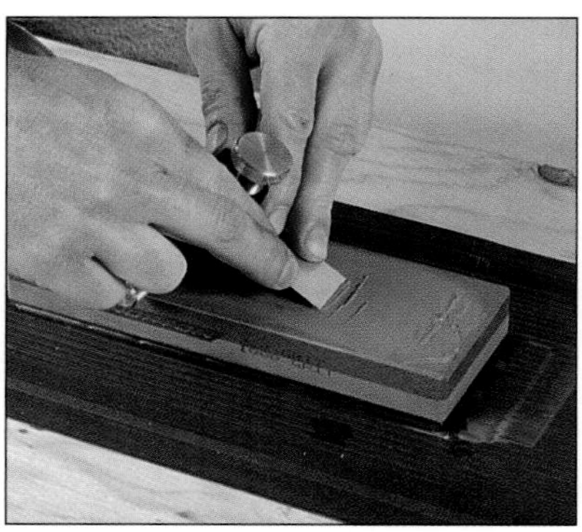

FIGURE D: Place the chisel and guide on the coarse side of the stone and lightly move the blade back and forth.

FIGURE E: Once a uniform bevel is created on the blade edge, turn over the stone and finish sharpening on the fine side.

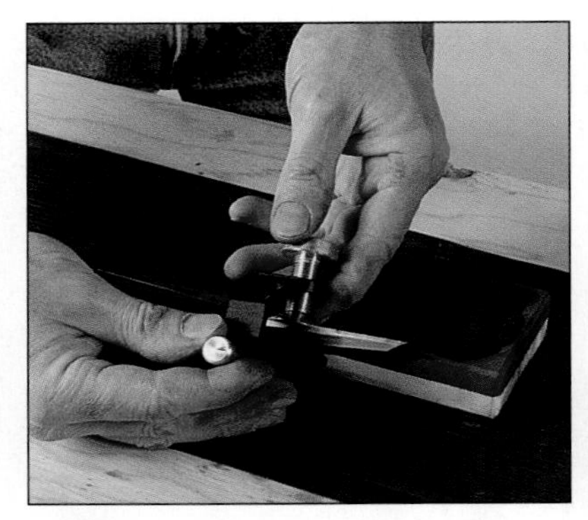

FIGURE F: To create a micro-bevel, adjust the knob on the side of the guide, then work the blade on the fine side of the stone.

FIGURE G: Flatten a worn stone by rubbing it across a piece of wet/dry sandpaper on top of a plate glass base.

6 When you're done with the bevel, remove the edge burr by lightly rubbing the back of the blade on the stone. Be sure to keep the blade flat. When you're satisfied with the edge, dry off the tool completely.

Secondary Bevel

1 If you want to hone a secondary bevel on the edge, simply adjust the honing guide—by turning its spring-loaded knob—for an angle that is 1° or 2° steeper than the primary bevel *(see Figure F).*

2 Move across the stone as before until you achieve a good edge like the one shown on the chisel on page 12. This secondary edge should extend back about 1⁄16 in. from the tip of the tool. Use the same techniques to sharpen a plane iron.

Flattening the Stone

As mentioned earlier, waterstones work well because they are soft. As a result, their surfaces wear quickly, which means that new cutting material is exposed almost continuously. Unfortunately, this softness does make them susceptible to gouging and uneven wear. Occasionally, you will have to flatten a stone to keep it performing well.

1 Take a piece of 1⁄4-in.-thick plate glass and lay a sheet of 220-grit wet/dry sandpaper on it, abrasive side up.

2 Wet the paper, then place the stone over it and rub the stone back and forth until any gouges or scratches disappear *(see Figure G).*

Text and Photos by Neal Barrett

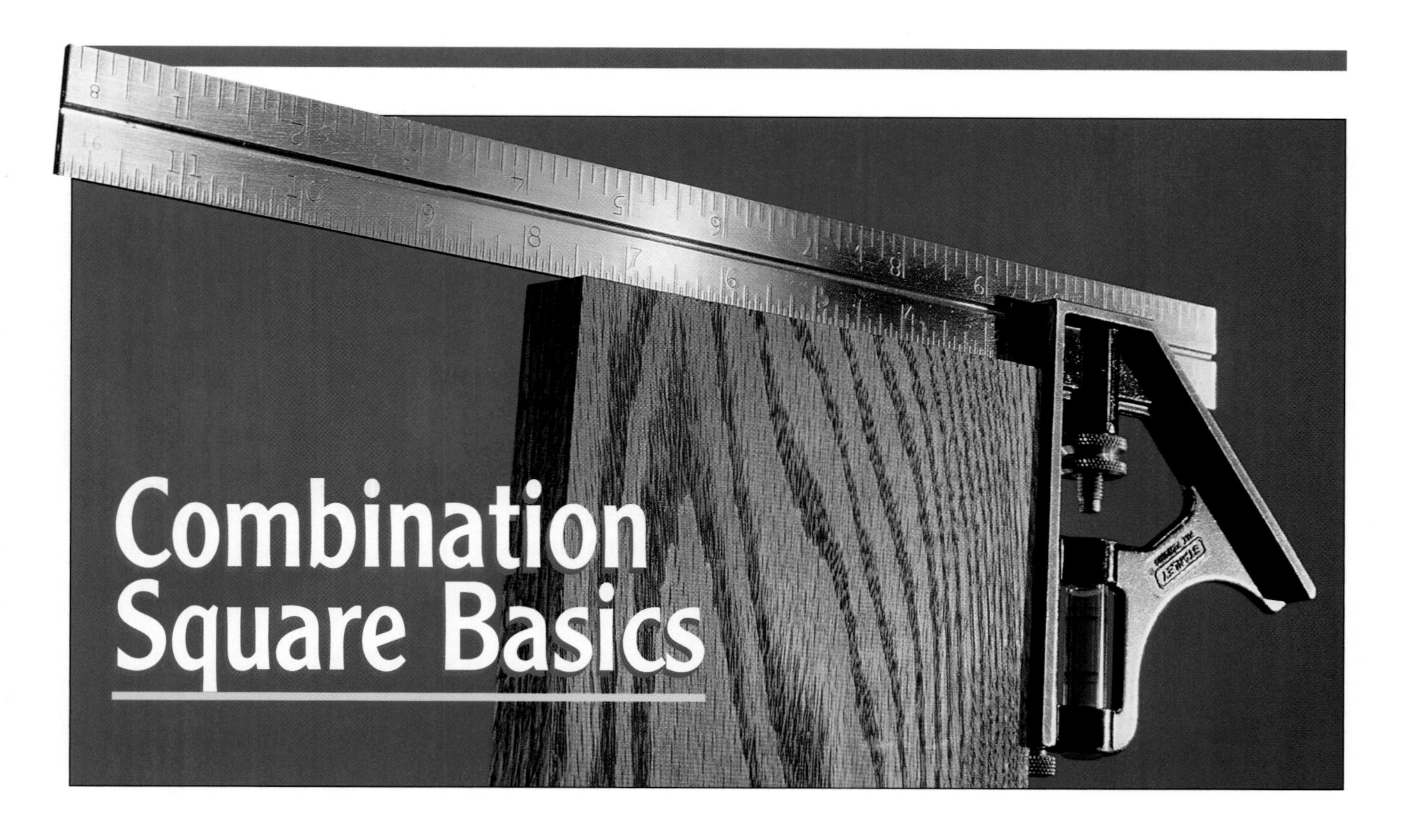

Combination Square Basics

If you work in wood, there's a good chance that most everything you'll ever build will be based on a 90° angle. And, if you expect to do precise work in both layout and assembly, you need a tool that accurately gauges 90°—you need a square.

While there are several specialized types available, including large framing squares, trysquares and precision steel squares, one of the most popular and useful is the combination square.

As its name implies, this tool does more than an ordinary trysquare. In addition to handling 90° angles, the combination square has a 45° shoulder for checking and laying out miters. Plus, by loosening a knob, you can reposition the square's head along the graduated, rule-type blade to suit the job. The combination square has a scriber housed in the head and there's also a vial for checking level and plumb *(see Figure A).*

Your square is the standard by which you cut and assemble your work. And, unless it's accurate, everything from basic joinery to the fit of drawers and doors will suffer. Fortunately, there's an easy way to test your tool.

Testing for Accuracy

1 Tape a sheet of paper to a board that has a perfectly straight edge.

2 Hold the square against the edge and draw a line along the outer edge of the blade.

3 Flip the square over so the opposite side of the blade faces up, align the square on the edge of the stock and draw a second line about 1⁄32 in. from the first *(see Figure B).* If the square is accurate, the lines will be parallel.

4 If you see any deviation, check to see that the edge of the head that abuts the work is perfectly flat. If it isn't, you may be able to correct the problem by dressing the surface on a piece of 100-grit sandpaper that rests on a flat surface such as plate glass *(see Figure C).* However, it often makes more sense to simply discard the square and buy a new one.

FIGURE A: The combination square does the job of an ordinary trysquare and more. A steel-rule-type blade is adjustable in the head. The head has a 45° shoulder for miters, and contains a scriber and level vial.

5 To check the 45° shoulder on the head, gauge it against a reliable 45° drafting triangle *(see Figure D)*. Be sure to check the triangle first by marking test lines from both of its faces as described for the square.

FIGURE B: Check your square by marking lines from each face of the blade. If lines are exactly parallel, the square is accurate.

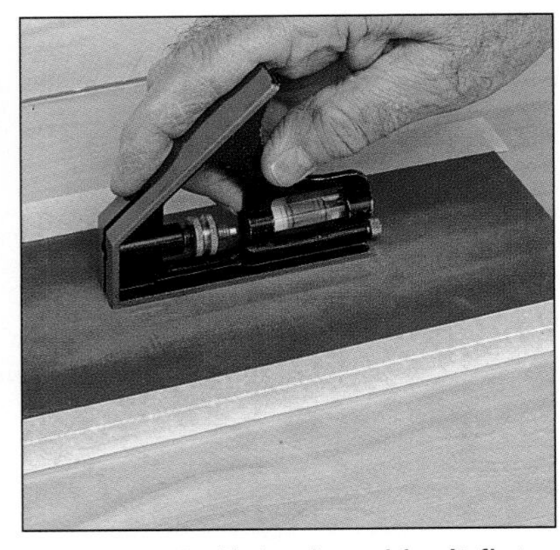

FIGURE C: If the head isn't flat, you may be able to repair it. True it on sandpaper backed by a flat surface such as plate glass.

Using the Square

1 To use a square to mark a line across a board, hold your pencil at the desired location and, with the head of the square against the edge of the work, slide the square to meet the pencil point *(see Figure E)*. Draw a line along the edge of the blade.

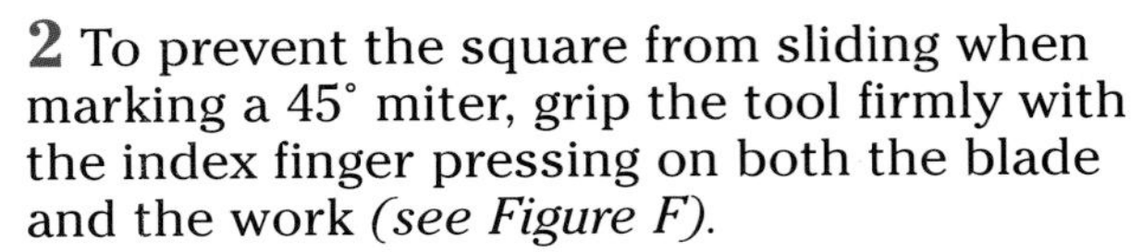

2 To prevent the square from sliding when marking a 45° miter, grip the tool firmly with the index finger pressing on both the blade and the work *(see Figure F)*.

3 When you need a line that's finer than the pencil line, use a knife or the scriber that's stored in the head of the square.

4 To check the end of a board, hold the square so the edge of the blade rests on the end of the workpiece *(see Figure G)*. Then, hold the work up to the light. If the stock isn't square, you'll see light under the blade on one side or the other. Use a similar system for checking the inside of framed corners and cases.

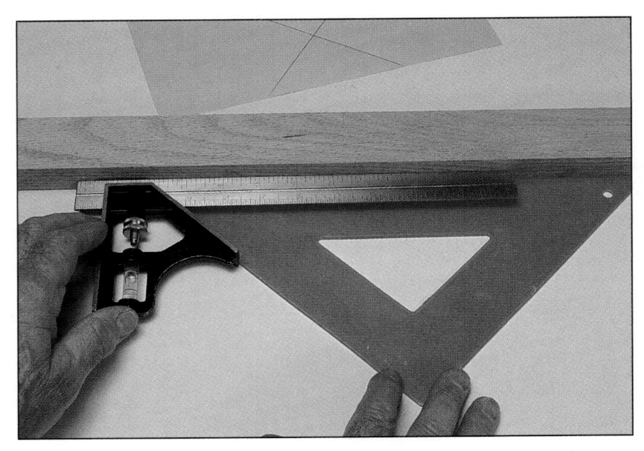

FIGURE D: Use a 45° drafting triangle to check your square's miter angle. Be sure to check the accuracy of the triangle first.

FIGURE E: To draw a precisely positioned square line, first position your pencil. Then, slide the square to the pencil and draw the line.

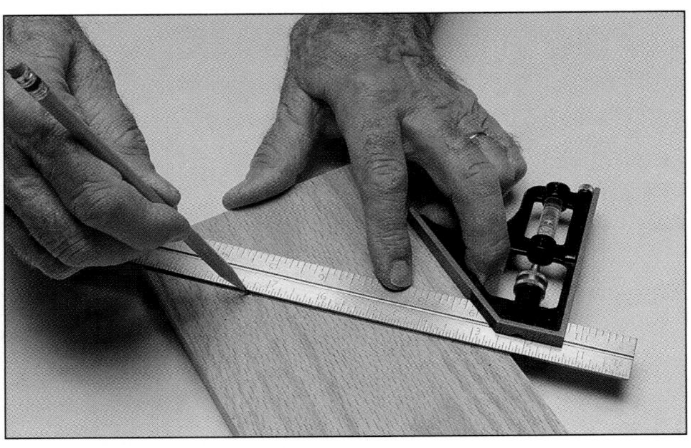

FIGURE F: When using the square to mark a miter, grip the tool firmly to prevent it from shifting. Use scriber or knife for a finer line.

FIGURE G: To check for square, hold the blade edge against the stock. Light visible between the blade and workpiece shows error.

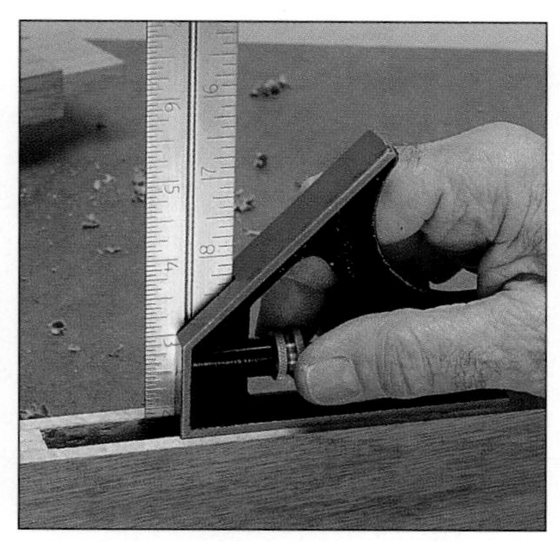

FIGURE H: To check the depth of a mortise, loosen the blade, rest the head on the work and slide the blade to the bottom.

FIGURE I: Use the square as a marking gauge by positioning the blade and holding a pencil at the end as you move the square.

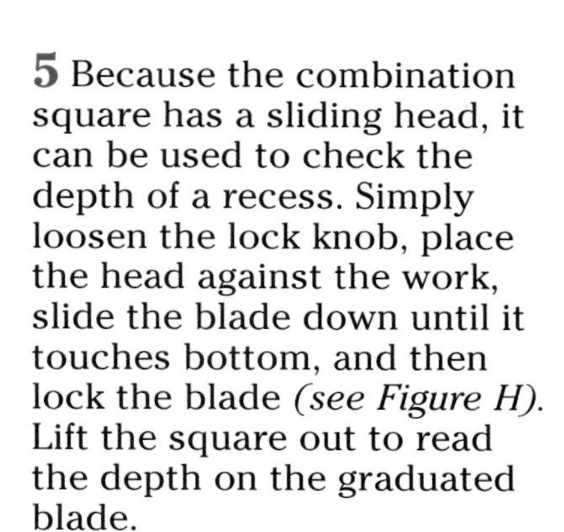

5 Because the combination square has a sliding head, it can be used to check the depth of a recess. Simply loosen the lock knob, place the head against the work, slide the blade down until it touches bottom, and then lock the blade *(see Figure H)*. Lift the square out to read the depth on the graduated blade.

6 The sliding head also enables the combination square to be used as a marking gauge. To draw a line parallel to an edge, hold your pencil against the end of the blade, then pull both the square and the pencil toward you *(see Figure I)*. To help hold the pencil in place, file a small V notch in the end of the blade.

FIGURE J: When you need a steel bench rule for layout or cutting, simply slide the head off the graduated steel blade.

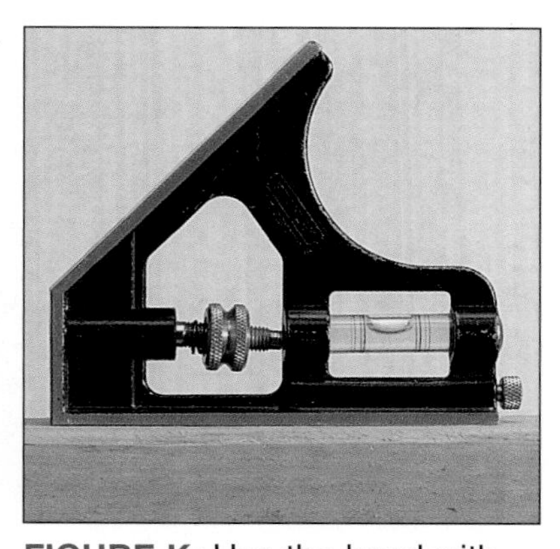

FIGURE K: Use the head with blade removed to check for level. Reinstall the blade and check vertical surfaces to see that they're plumb.

7 When you need a steel rule for layout work or cutting with a utility knife, simply slide the blade out of the head *(see Figure J)*. The blade on a typical combination square has scales with 1/32-, 1/16- and 1/8-in. gradations.

8 With the blade removed the head with integral vial is used to check for level *(see Figure K)*.

9 To check for plumb, reinstall the blade and hold its edge against a vertical surface.

Text and Photos by Rosario Capotosto

Bench Plane Basics

Most well-equipped workshops would be hard pressed to get a day's work done without a modern thickness planer and jointer. However, before these high-powered surfacing machines arrived on the scene, skilled artisans got the same work done—they just did it all with hand planes.

While there are many types of planes, the most generally useful is the bench plane. Today's metal bench planes are based on the Bailey pattern developed over a century ago. They're available in several sizes that include the 9-in. smooth plane, 14-in. jack plane, 18-in. fore plane and 22-in. jointer plane. The longer the plane, the more effective it is at flattening, or truing, a surface. The shorter planes are used for final smoothing or preliminary roughing out. If you're going to own only one plane, the jack is a good choice.

Adjusting the Plane

1 The plane iron is mounted on a 45° bed called a frog. At the back of the frog is a depth adjustment knob and a lateral adjustment lever. The frog itself can be adjusted forward or backward to alter the size of the opening in front of the cutter, or mouth *(see Figure A)*. Set the mouth wide for coarse work and narrow for finish planing.

2 To stiffen the beveled cutting edge of a bench plane, the cutter features a cap iron that's clamped in place with a single screw. The cap iron is shaped to exert pressure on the plane iron. In addition to helping to reduce the tendency to chatter, the cap iron

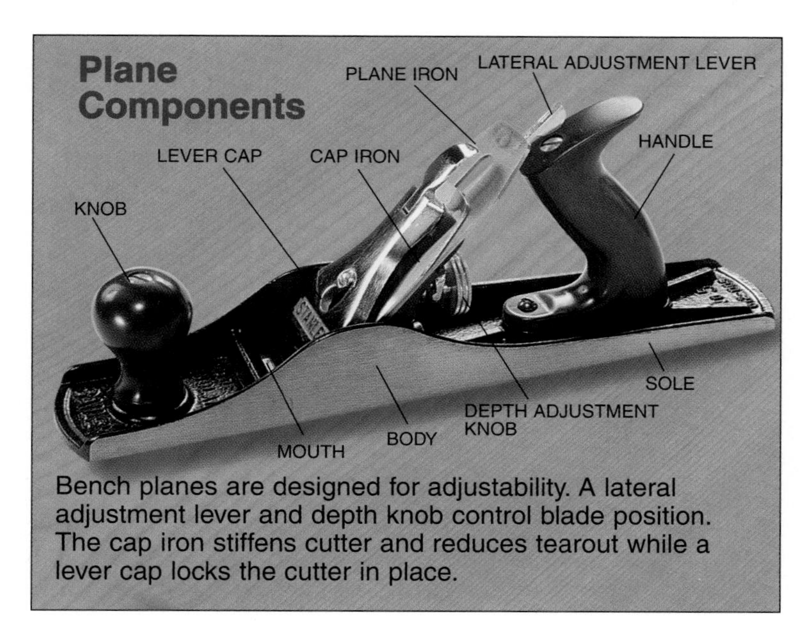

Bench planes are designed for adjustability. A lateral adjustment lever and depth knob control blade position. The cap iron stiffens cutter and reduces tearout while a lever cap locks the cutter in place.

also curls and breaks the shavings as they're produced. This helps when it's necessary to plane against the grain. Set the cap iron to about 1⁄32 in. from the cutting edge for fine planing, and at about 1⁄16 in. for general work. For very coarse work, move it back to about 1⁄8 in. from the cutting edge *(see Figure B).*

3 Rotate the depth adjustment knob to adjust the blade's projection beyond the sole. Sight down the sole to check cutting depth and to ensure that the cutting edge is parallel to the sole. If necessary, move the lateral adjustment lever to the right or left to align the cutter with the sole *(see Figure C).*

Sharpening the Plane Iron

Regardless of how expertly you handle the plane, good work will only follow if you keep the plane iron sharp.

1 Use a honing guide that holds the iron at the correct angle. It will take only a few minutes to properly hone the edge of the iron.

2 Use a fine-grit oil or water stone to hone the bevel.

3 Remove the cutter from the guide and hone the back to remove the wire edge.

4 Follow with a few strokes on a very fine polishing stone or leather strop.

Jointing an Edge

1 To joint an edge straight, square and smooth, first orient the stock so you're planing with the grain—the grain appears to move uphill in relation to the planing direction.

2 Apply candle wax to the plane sole to reduce friction.

3 Firmly grip the knob and handle and make full, continuous passes.

4 When making the final passes to square the edge, reposition your front hand to grip the plane body while your fingers ride against the wood. In this way your front hand helps to stabilize the plane and hold it square *(see Figure D).*

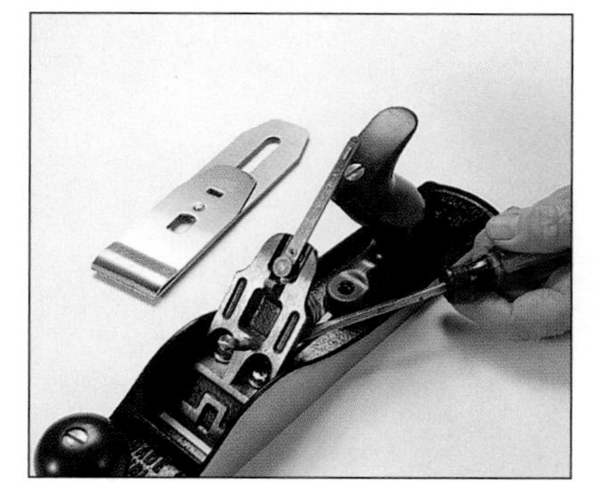

FIGURE A: With the cutter removed, the plane bed, or frog, is accessible. Adjust forward or backward to alter mouth size.

FIGURE B: Adjust the curved cap iron so it's about 1/16 in. behind the cutting edge for general work. Bring the cap iron closer for fine work.

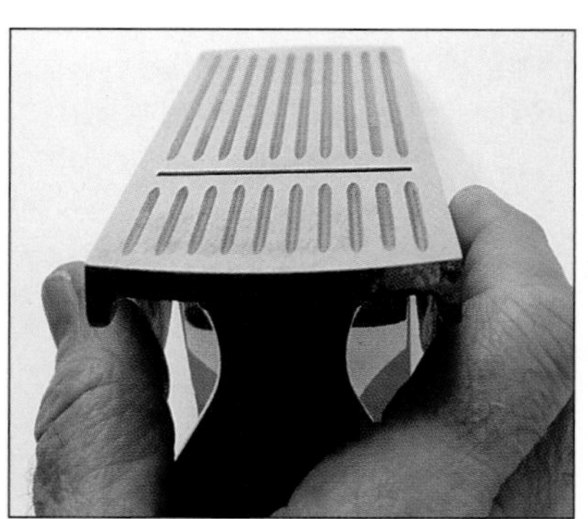

FIGURE C: Sight down the sole of the plane to check for alignment and cutting depth. The corrugated sole on this plane reduces friction.

FIGURE D: When jointing an edge square, place your front hand on the plane body so your fingers can help guide plane along the stock edge.

FIGURE E: Use a shooting board to quickly plane square edges. Hold the plane side flat against the lower board when cutting.

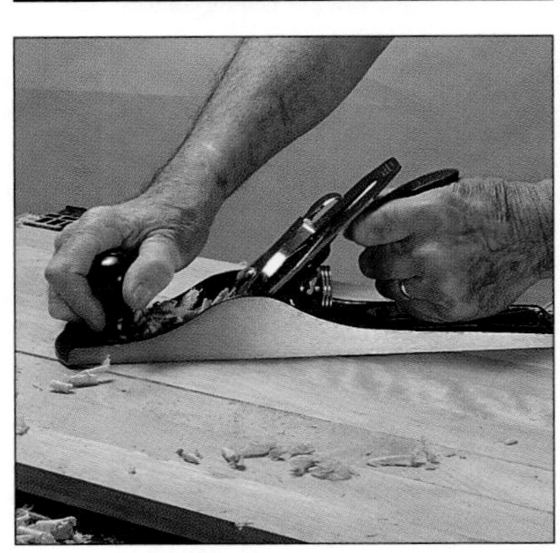

FIGURE F: Plane diagonally across a rough panel to quickly flatten the surface. Make the finish cuts along the grain with a finely set cutter.

FIGURE G: To plane uniform chamfers and bevels, orient the workpiece at the appropriate angle and hold the plane level.

Jointing Using a Shooting Board

As an alternative to squaring the workpiece edge by feel, construct a shooting board to hold the work and guide the plane. Clamp your work to a flat, straight board with a spacer board in between *(see Figure E)*. Lay the plane on its side on the first board and plane the edge of the workpiece.

Truing a Panel

1 When truing the surface of a panel, set the plane iron for a shallow cut. If you're doing a lot of coarse surfacing, have a second plane on hand with the cutting edge ground to a slightly convex profile to handle the rough planing.

2 Make passes across the grain, then diagonally across the workpiece *(see Figure F)* and finally parallel with the grain.

3 Use a pair of straight sticks, called winding sticks, to check that the board has no twist. Lay a stick at each end and sight down the work to make sure the sticks are parallel.

4 Finish leveling the surface with fine cuts along the grain.

5 To bring the panel to desired thickness, use a marking gauge held against the dressed side to mark the finished thickness on the edges and ends.

6 Plane down the rough side until the scored line from the marking gauge is split.

7 When edge-gluing narrow boards to make a wide panel, arrange the stock so the grain direction is the same on each. Otherwise, you'll encounter reversing grain at the joint line in the assembled panel.

8 It's good practice to develop a feel for holding the plane level throughout its stroke. When planing chamfers and bevels, maintain this level attitude by repositioning the work instead of the plane *(see Figure G)*.

Text and Photos by Rosario Capotosto

18-volt Cordless Drills

Remember when we were happy with those old 9.6-volt cordless drills? Sure, you still needed your plug-in tool for the tough stuff, but, face it, cutting the cord was one of the best ideas to come down the pike in quite a while. Then, just as we were wondering where to buy fresh battery packs, 12-volts came along. Now here was power—enough torque to drive serious deck screws and even lags if we didn't push it. Buy new 9.6-volt battery packs? Heck, might as well get a new 12-volt drill.

By then, we were hooked. And like the classic addict, we needed a bigger hit each time. Sure, some of us jumped at the 14.4-volt teasers, but down deep, we all knew what we were waiting for—the big ones—the 18-volt go-anywhere, do-anything drill/drivers. Then we'd be happy.

Well, here they are. But before you dip into next month's grocery money to buy one, take some time to evaluate your needs against what these tools offer. There's no question that they're big and powerful. On the other hand, an ordinary ⅜-in. corded drill will blow the best of them out of the water. And then, you just might want to bide your time. DeWalt has already introduced a 24-volt drill and a big Bosch model isn't far behind.

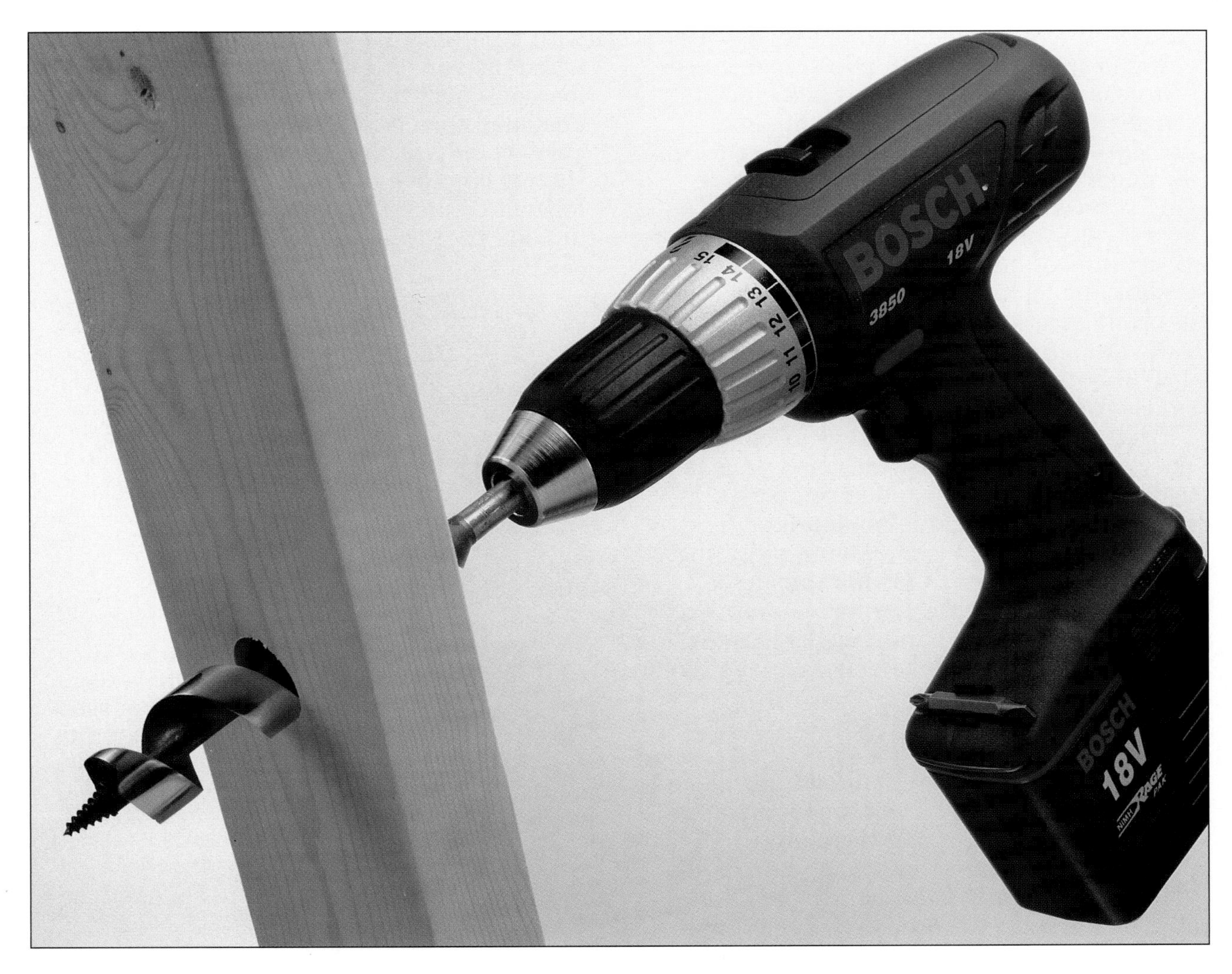

18-Volt Variable-Speed Reversible Drill/Driver Specifications

Manufacturer	Model	Price[1] (MSRP/street)	Ampere-hours (type, no. included)	Maximum rpm (low, high range)	Chuck size (type)[2]	Maximum torque (in.-lb.)	Clutch positions[3]	Weight (pounds)	Accessories/ features[4]	Charger
Black & Decker	FS18	$149/ $130	1.7 (NiCd,2)	400, 1300	3/8" (one hand)	266	24	4.6		3-hour
Bosch	3850	$396/ $305	2.0 (NiMH,2)	500, 1650	1/2" (one hand)	425	16	4.8	C	1-hour
Craftsman	27199	$225	2.0 (NiCd,2)	400, 1400	1/2" (two hand)	345	24	6.3	A, B, D, E, F	1-hour
DeWalt	DW995K-2	$494/ $269	1.7 (NiCd,2)	450, 1450	1/2" (two hand)	355	24	5.6	C	1-hour
Makita	6343DWAE	$458/ $260	2.0 (NiCd,2)[5]	450, 1400	1/2" (two hand)	404	17	5.5	C, E, G	1-hour
Milwaukee	0521-20[6]	$440/ $260	2.0 (NiCd,1)	450, 1500	1/2" (two hand)	400	20	5.6		1-hour
Ryobi	CTH1802K	$175/ $169	1.7 (NiCd,2)	350, 1300	1/2" (two hand)	300	24	4.3	C, D	1-hour
Skil	2882:04	$248/ $130	1.3 (NiCd,2)	400, 1200	1/2" (two hand)	200	16	4.5	C, F	3-hour[7]

1. Price as suggested by the manufacturer. **2.** One-hand or two-hand locking—all chucks are keyless. **3.** Including drill mode (locked). **4.** All drills come with a carrying case. **5.** 2.2-Ahr NiMH battery pack available. **6.** Midhandle model is 0522-20. **7.** Kit 2882:01 includes 1-hour charger.
A. slot driver bit. **B.** Phillips driver bit. **C.** Phillips/slot combination bit. **D.** wrist strap. **E.** auxiliary handle. **F.** built-in bubble level. **G.** accessible brushes.

Basic Features

All of the tools we tested are reversible, variable-speed drills with two speed ranges. Cordless drills need low- and high-speed ranges to make the best use of minimal power—the low range provides more torque for auger bits and more control when driving screws, while the high-speed range is best for boring small holes.

In terms of design, all of our test tools except one feature a well-balanced, midhandle configuration that reduces fatigue in awkward situations. Milwaukee offers its drill in both midhandle and pistol-grip versions—we tested the pistol grip.

Each tool comes with a keyless chuck. The standard design features two rings, and tightening it is a two-handed operation. In situations where the bit jams in the workpiece, these chucks can lock up, requiring a couple of wrenches to loosen the grip. Bosch and Black & Decker have equipped their drills with one-hand chucks that, in our testing, were better because they didn't lock up.

Each drill also has a clutch—a graduated ring behind the chuck that's intended to disengage the motor at predefined torques and, thereby, automatically set threaded fasteners appropriately. Setting the clutch, though, is an exercise in trial and error. If you're driving only a few screws in a variable material like wood, you may end up with a better jobif you simply stop the drill when it's gone far enough. And finally, while it's not a necessity, all of our tools are designed to accept a strap—a useful accessory when you have to let go of the drill but there's no place to set it down. Only the folks at Craftsman and Ryobi actually include the strap, though.

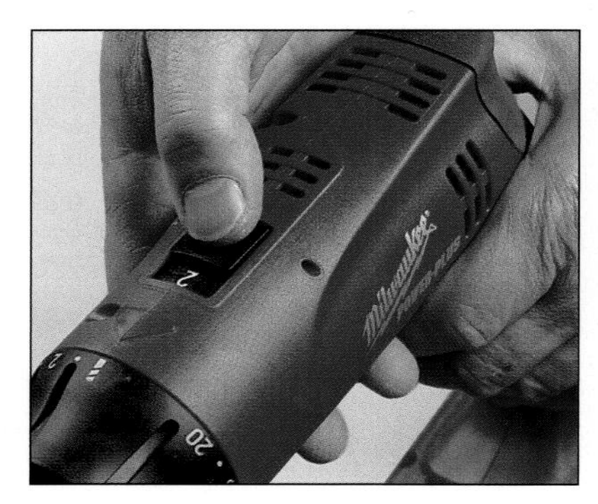

The drills have a speed-range selection switch on the top of the housing. Low range is best for high-torque operations.

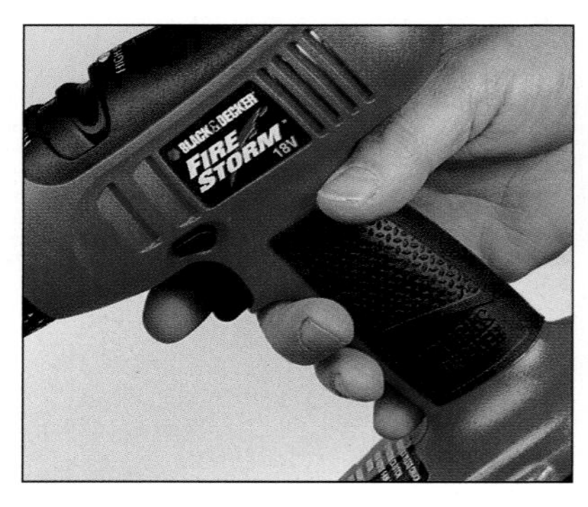

A sliding forward/reverse switch is directly above the trigger. A nonslip rear grip is found on all except Makita and Ryobi.

Just Charge It

Today's 1-hour chargers do more than just fill your battery pack with juice. They also make sure the pack is within an appropriate temperature range for charging, determine whether the batteries are defective and drop down into a trickle-charge mode after charging is complete. While most chargers keep you informed as to what's going on, some designs are better than others in terms of getting the message across.

The Makita charger, for example, has a light that blinks or shines steadily, in red or green, to indicate charging mode and battery condition. Each situation has a corresponding light code, and a clear label on the charger indicates their meanings.

In contrast, the Bosch charger has a single red light that flashes during charging and stays lit when charging is complete, or if the battery is too hot or cold—a bit more confusing. There's no indicator for a defective battery.

Like the Makita charger, the Milwaukee offers an expressive array of lights and graphics. The DeWalt has a single red light with its own Morse-code-style language to let you know what it's doing.

Ryobi and Sears both have a readily understandable array of colored lights to indicate charge status.

In terms of user-friendly information, the most deficient chargers are the 3-hour units from Black & Decker and Skil. Both have a single red light that simply indicates that the charger is plugged in and on. It's up to you to set your alarm clock for 3 hours and hope the charging is done when the bell goes off.

The friendlier chargers, like the Makita (left), offer lights that describe charging progress. The Skil and B&D versions simply indicate when the charger is plugged in. The minimal Skil charger (top) consists of a cap wired to a plug-in power unit.

Battery Choices

Typically, cordless drills are powered by nickel-cadmium (NiCd) battery packs. Over the past year there has been a wave of interest in nickel-metal-hydride (NiMH) units. Although toolmakers claim these new batteries can deliver more power, their introduction may have as much to do with regulations on battery disposal. The cadmium in NiCd cells is extremely toxic, while NiMH batteries can be tossed with the rest of the garbage.

The amount of electricity in a battery pack is rated in terms of ampere-hours (Ahr). The most potent NiCd battery packs are rated at 2.0 Ahr, while NiMH packs of up to 3.0 Ahr are in the works. However, these powerful NiMH batteries may not have the recharge cycle life we've come to expect with NiCd.

As of now, NiMH is yet to be embraced by all cordless manufacturers. In fact, some are quietly stepping back from their initial enthusiasm. In our group, only Bosch sells its drill with NiMH. Makita offers NiMH as an accessory, but sells the drill with NiCd.

Black & Decker FS18

If you want 18 volts but balk at the big price tags, the B&D Firestorm may suit your needs. It has all the basic features of a good cordless drill, plus a convenient one-hand chuck. What you sacrifice is power and endurance.

You'll also sacrifice 1-hour charging—a feature common to all of the other tools except the Skil. The B&D charger takes about 3 hours, and the small red indicator light tells you only that the charger's on—not that the battery is charging or at full charge. The Firestorm was the only tool with a ⅜-in. chuck capacity—all the others handle ½-in. bits.

On the good side, the tool is well-balanced with a comfortable rubber grip. The 23-position clutch offers a uniform range of torques.

Bosch 3850

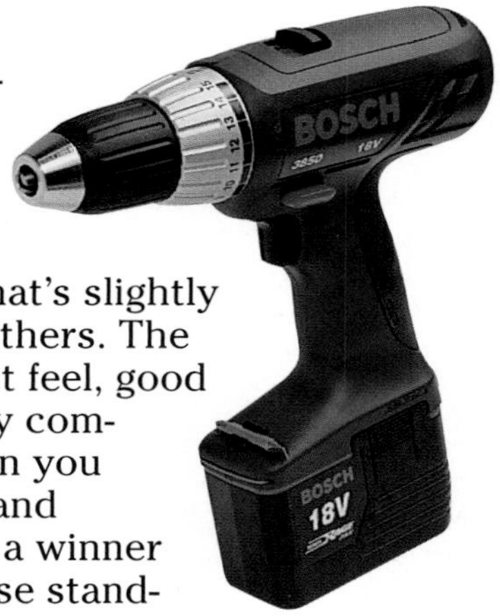

The Bosch engineers have taken advantage of NiMH's higher energy density to create a battery that's slightly smaller than the others. The tool has a compact feel, good balance and a very comfortable grip. When you factor in its one-hand chuck, this tool is a winner from the ease-of-use standpoint.

While the manufacturer's specs place the Bosch in the lead in terms of torque, our sense of the tool's power places it just below the Makita and Milwaukee tools. In terms of high-torque lag-driving endurance, the Bosch ranked in the middle—about on par with the Ryobi and DeWalt tools. However, tests with auger and spade bits place the Bosch at the head of the pack. The 15-position clutch progresses uniformly through its torque range.

Craftsman 27199

The entry from Sears is the largest tool in the group, measuring 12 in. in length and weighing in at 6.3 pounds. In addition to the two battery packs, charger and driving bits, it comes with a wrist strap, auxiliary handle that mounts behind the chuck and a built-in bubble level on the back of the housing.

While the extra size and weight make the Craftsman drill a little harder to handle, we found the tool had very good power. It ranked near the top for lag bolt-driving endurance and near the middle of the group when boring 1-in. holes.

DeWalt DW995K-2

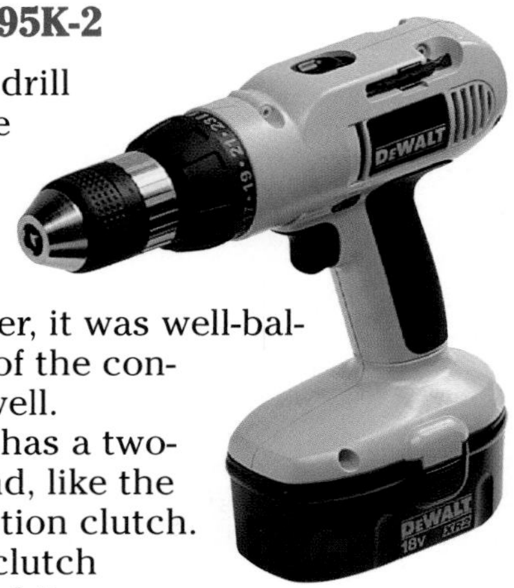

The DeWalt drill qualifies as the winner in the no-surprises category. The tool had good power, it was well-balanced and all of the controls worked well.

The DeWalt has a two-hand chuck and, like the B&D, a 23-position clutch. We found the clutch torque range a bit more erratic than the B&D's—settings 1 through 7 progressively sank a 2½-in. screw deeper and deeper, while the next six settings left the screw nearly flush with the surface of the wood.

DeWalt equips its tool with a modest 1.7-Ahr battery pack, and it rates the tool at 355 in.-lb. Both specs are on the short side when compared with other tools in DeWalt's price niche. Our endurance tests place the NiCd-powered tool in the middle of the pack, but our impression is that the drill easily handles most driving and boring chores. While the 1.7-Ahr pack may not have the staying power, you may get more charges out of it than from a more potent unit.

Makita 6343DWAE

When you're paying over $250 for a cordless drill, you have a right to expect performance—and a few extras, as well. Here, the folks at Makita seem to be on your side. We found this unit running a close second to Bosch in terms of run time per charge. However, when it comes to power, the Makita simply feels like it has more than enough for whatever job is at hand.

Like Craftsman, Makita has thoughtfully provided a handle to help control the ample 404 in.-lb. of torque. However, Makita goes one step further by providing a depth-stop rod for the handle. Makita is also the only tool with brush access covers. While it comes with 2.0-Ahr NiCd batteries, you can get 2.2-Ahr NiMH, and there are plans to offer 3.0-Ahr 18-volt packs.

Milwaukee 0521-20

Milwaukee offers 18-volt drill/drivers in both midhandle and pistol-grip configurations. In our opinion, if you're boring big holes, the pistol grip works fine. Switch to driving a box of screws, and you might be happier with a midhandle.

On all counts, the Milwaukee entry is a heavy-duty performer. In terms of endurance we rank it with the Makita, right under the Bosch. While its rated maximum torque falls below the Bosch's, our impression is that it's a more powerful tool.

While every other battery pack slides up into the handle, Milwaukee's slides across the bottom of the grip. And the pack design is offset—sliding it in from the front moves the drill's balance point forward, while sliding it in from the rear shifts the balance toward the grip. About the only thing this package lacks is a second battery pack.

Ryobi CTH1802K

If you're frustrated by the high prices of the so-called pro drills but demand more than what the consumer lines have to offer, this Ryobi unit may be the answer. Equipped with a 1.7-Ahr battery pack and a maximum torque of 300 in.-lb., the CTH1802K has acceptable qualifications, and our tests place it just below the DeWalt in terms of performance and battery endurance. When you factor in the tool's light weight, comfortable grip and user-friendly charger, it's hard to beat—especially with an estimated street price of about $100 less than the big guns'.

The keyless chuck on Bosch and B&D drills can be conveniently tightened with one hand. Bosch drill shown.

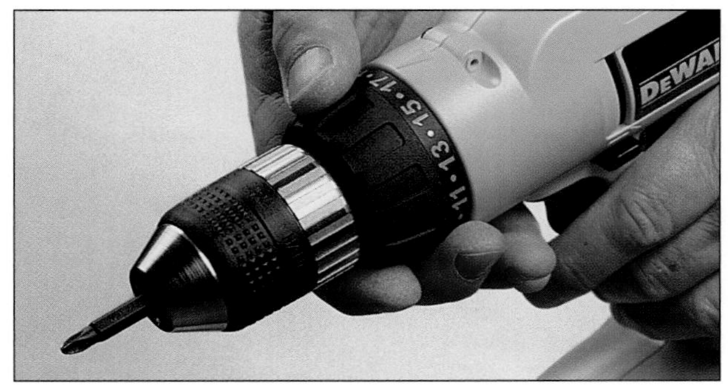

All drills feature a clutch that's numbered to indicate relative torque at which the motor disengages.

Skil 2882:04

Like the B&D Firestorm, Skil's Warrior 18-volt drill is an entry-level tool, designed to supply 18-volt performance at a street price of about half what the pro tools sell for. However, when pushed to the limit boring big holes and driving heavy-lag screws, we found the tool's performance right in line with the modest price. This tool has the lowest maximum torque (200 in.-lb.) and battery Ahr (1.3) ratings of the tools in our group.

On the plus side, this Skil cordless drill is light in weight and fairly comfortable to hold, and the controls are easy to operate. While the kit we tested comes with a 3-hour charger, the drill is also available with a 1-hour version.

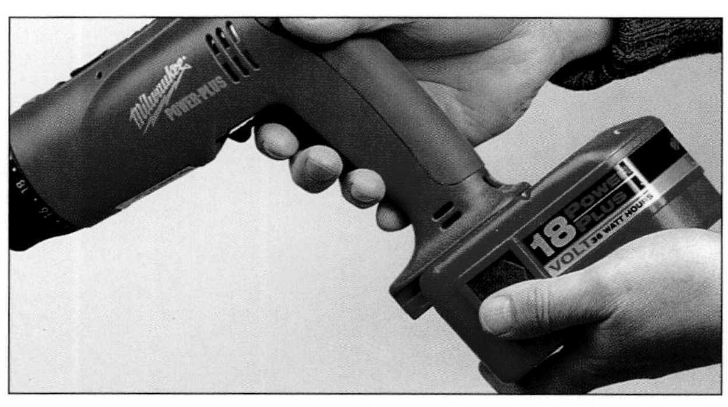

Unlike most drills, where battery packs fit up in the handle, the Milwaukee has a pack that slides on from either direction.

Makita's drill is the only one with brush access covers. It's easy to check commutator wear and replace brushes.

Decisions, Decisions

If you're in the market for a cordless drill, first estimate whether a 12- or 14.4-volt model will do the work. If you're only driving drywall screws and boring the occasional 1-in. hole, dropping your voltage requirements may put a higher-quality drill within your budget.

If you opt for 18 volts and money isn't an issue, you might as well get the most punch that you can. Here, our choices are the Bosch, Makita and Milwaukee drills.

Right behind these are the Craftsman and DeWalt drills. The DeWalt comes with a reputation for quality and service, but the Craftsman's a little cheaper and comes with more features.

The performance-per-dollar winner is, of course, the Ryobi. The Skil and B&D entries both offer more performance than 12-volt versions, but don't come near the other tools in the 18-volt group.

Text and Photos by Thomas Klenck

Wood Glue

If there's one material, besides wood, that's central to furniture making, it's wood glue. Since ancient times, glue has been used to assemble furniture and it's not difficult to see glued pieces that are hundreds of years old. Just take a trip to an art museum and look at furniture from ancient Egypt or the European Renaissance. While the purpose of glue hasn't changed over the years, the technology certainly has. Now there are any number of specialty glues designed for all sorts of different applications. Fortunately, only a few play an important role in furniture making.

Hide Glue, Epoxies and Polyurethanes

The earliest glues were hide glues and these are still in use today. Hide glue is made from animal products and it's extremely useful for projects, like musical instruments, that often require disassembly to make repairs. Because heat and humidity cause hide glue to release its bond, it's a relatively simple matter to separate pieces without damaging them. Hide glue also cures slowly, so it can be a good option for difficult joints or constructions that take a long time to assemble. However, the goal of most furniture work is to create something that can withstand exposure to just these elements, heat and humidity. Fortunately, today's furniture maker has other options.

Two-part epoxies are probably the most durable of all adhesives. For situations where extreme water resistance is required, epoxy is the best choice. Unfortunately, it's pretty difficult and messy to use. And epoxies are quite toxic, so you need to don gloves and a respirator to protect yourself from chemical exposure. These inconveniences make epoxies a bad choice for everyday work.

One of the latest adhesives to appear on the furniture making scene is polyurethane glue, which is supposed to be well-suited to just about any gluing job. This glue performs unlike any other. It actually cures by being exposed to moisture, so it's a good choice when moisture resistance is an issue. You even have to dampen wood surfaces before applying this glue. This product changes into a foamlike substance as it cures and in the process expands out of the joint. This can make sanding away the glue more difficult. Also, because it's so new, it can't claim the long-term successful track record that other glues enjoy.

White Glues and Yellow Glues

The most common furniture making glues are polyvinyl acetate adhesives, known casually as white and yellow glues. While white glue is a good general adhesive that can be used on most porous materials, yellow glue has been specifically formulated for interior woodworking applications. Yellow glue is usually referred to as aliphatic resin glue.

Neither of these glues work well if a water-resistant bond is required. They are poor choices for building things like high-end exterior doors or outdoor furniture. For these purposes there are water-resistant formulations of yellow glue. These are technically known as cross-linking PVAs, and they cure through chemical reaction, instead of evaporation. For general woodworking use, this glue is interchangeable with normal yellow glue except that it can't be cleaned up with water after it cures.

While each of these products has its place in the furniture maker's repertoire, aliphatic resin glue is the best choice for the beginner. It's easy to use, requires no mixing, is nontoxic and cleans up with water. It also sands cleanly, without overclogging the sandpaper, and leaves an invisible glue line if the joint is tight. Like all glues, however, it does have a finite shelf life. Once it's open, it's only good for about a year. If you notice that the glue starts to smell sour and becomes extremely thick or stringy, it's past its prime and shouldn't be used.

Gluing Techniques

In preparing a glue joint, it's important to keep a couple of ground rules in mind. First, while modern glues are amazingly strong, if joints do not fit properly or the glue is not allowed to cure correctly, the bond will almost surely fail. And second, less is more. A thin, even layer of glue will form a strong bond between two pieces of wood, but a thick cushion of glue does just the opposite. It weakens the joint. And, usually you won't know if this joint is weak until the piece is done and has been used for a while.

To achieve a successful edge joint, the long mating surfaces must be perfectly tight all along their length. You shouldn't rely on clamps to pull bowed boards together because this places too much stress on the joint and eventually it will fail.

The fit of a mortise-and-tenon joint should also be precise, neither too tight nor too loose. If the parts must be forced together, there will be no room for the glue between the pieces and the joint will be starved. If, on the other hand, there is too much play in the joint, the glue layer will be too thick to perform properly when the piece is placed under stress. In a perfect joint, the glue layer is less than the thickness of a sheet of notebook paper.

In order to guarantee that there is sufficient glue in a joint, spread a thin layer on both mating surfaces. In a mortise-and-tenon joint, this means coating both the mortise and tenon walls with glue. You can use a scrap stick, a small brush or a narrow roller to do this. Just make sure the coverage is complete and even.

The mating surfaces must also be clean, dry and free of contamination before you spread the glue. Oil, waxes and some chemicals, like silicone, will resist the glue. And dust and water can cause the joint to fail by preventing good contact between wood sur-

The basic furniture maker's glue assortment includes (from left) waterproof yellow glue, white glue, yellow glue, epoxy (in front), polyurethane glue and hide glue.

faces, or diluting the strength of the glue.

All glues have a recommended open time, which defines the amount of time you can safely leave the glue exposed to the air before assembling the joint. For most yellow glues, this is about 10 minutes. But the open time will vary with the temperature and humidity of your workroom. Hot, dry conditions will cause the glue to set quicker. In complicated assemblies, where many joints must be prepared at once, it's important to factor the open time limit into the process. Sometimes, you'll have to assemble a project by making smaller subassemblies first. Also, keep in mind that yellow glue will not perform well in cold conditions. Most manufacturers recommend that both the room and the wood surfaces be at least 55°F before applying glue.

Once a joint is assembled, it must be clamped together. Clamps serve two purposes. They pull a joint together tightly and hold it in a fixed position while the glue sets. You should not apply tremendous force with your clamps because this will drive too much glue from the joint. Just firmly tighten them and set the assembly aside.

If a proper amount of glue has been spread and a proper amount of force used to tighten the clamps, you should see small beads of glue squeezing out of both sides of the joint. To remove this squeeze-out, allow it to set for about 20 minutes, and then use an old chisel or putty knife to scrape off the excess.

Some people recommend wiping the excess glue off with a damp rag. But this technique should be avoided because it can force glue into the surrounding wood pores—especially with open-grain woods. Unfortunately, this glue will not be apparent until you apply the finish, when it's too late to do anything easily to fix it. Finally, yellow glue should be allowed to set for at least an hour before you remove the clamps. And a full cure takes at least 24 hours, so don't disturb the assembly until this time has passed.

Text and Photos by Neal Barrett

Stock Selection

The raw materials for most furniture making projects fall into three general categories: softwood lumber, hardwood lumber and manufactured panels. The type of material you use for any given project depends on various factors: strength, hardness, grain characteristics, cost, stability, weight, color, durability and availability. Most beginning woodworkers have their first experience with softwood, usually pine lumber. It's soft and easy to work, and you don't need expensive shop equipment to get good results. It's also readily available at local lumberyards and home centers. But it does have some notable liabilities, especially if you're planning to make a fine piece of furniture. So before you buy a bunch of pine for that coffee table you have in mind, investigate all your options.

Softwood Lumber

Softwood is the wood from an evergreen or coniferous (cone-bearing) tree. Examples are the many varieties of pine, fir, spruce, hemlock, cedar and redwood. Some softwoods are used as veneer in plywood panels. But most of us are more familiar with this material in solid lumber form. Most of the wood at lumberyards and home centers falls into the softwood category. These woods are the mainstay of the home construction industry. They're used for framing lumber, window sash material, millwork stock and much more. Generally speaking, framing lumber, such as fir, spruce and hemlock, is not used for fine furniture making. But pine boards do play a significant role in cabinetmaking and architectural trimwork.

Pine boards are sold in nominal widths from 2 in. to 12 in. The actual measurements of these boards are usually ½ in. to ¾ in. less than their nominal dimensions. A similar convention applies to a board's thickness. A nominal 1-in. board is actually ¾ in. thick and a 5/4 board is actually 1$\frac{1}{16}$ in. thick. All softwoods are sold in even foot lengths from 8 ft. to 16 ft.

One of the advantages of using pine for a project is that you can confidently anticipate the sizes of lumber that are available and easily calculate the yield you can get from the stock. This can be much more difficult when you work with hardwoods—more on this later.

Unfortunately, pine and most other softwoods are less stable than most hardwoods. They absorb and lose moisture more readily and are normally

sold with a higher moisture content—up to 15%—than hardwoods, which usually have ranges from 6% to 8%. If you plan to use pine, purchase the lumber at least two weeks before starting your project and keep it indoors. Be sure to place stickers (small pieces of scrap wood) between the boards to permit good air circulation around each piece of stock. These procedures will allow the wood to reach an equilibrium with the indoor environment before you begin, which reduces the likelihood of dramatic wood movement after a project is complete.

While several grades of pine boards are routinely milled, in practice you'll find only two grades at most suppliers: select and No. 2 common. The common grade allows tight, solid knots in the face of the board and is most often used for paneling, shelving and paint-grade work. For furniture applications the select grade is the better choice. This stock is free from most knots, though tight pin knots and small resin pockets are allowed.

Pine is best used for designs where an informal or rustic appearance is preferred. The characteristic softness of the material, as well as its broad grain pattern, are well suited to furniture in Early American country styles.

Clear, oil-based finishes give pine an attractive warm amber cast. But applying stain, particularly oil-based ones, can present problems. Pine is extremely resinous and has a grain that can change dramatically from one area to the next on a given board. Consequently, the wood is likely to absorb stain in an uneven manner, creating a blotchy appearance. You can achieve a more uniform stained surface by applying a wood conditioner to the piece first. The conditioner limits the absorption of the stain. But keep in mind that using a conditioner isn't always successful. So, it's a good idea to run a test first on an inconspicuous part of your project to make sure you're happy with the results. Of course, pine is a perfectly good choice when a paint-grade surface is desired. The most striking characteristic of pine, however, is its softness. You can often leave an impression by simply running a fingernail down the board.

Hardwood Lumber

Hardwood lumber comes from deciduous trees, the ones that shed their leaves annually. Some popular domestic species are oak, maple, cherry, birch, walnut, ash and poplar. Of these common native hardwoods, only red oak and poplar are usually stocked in home centers and lumberyards, where they're frequently sold in the same sizes as pine boards. But most hardwoods are carried by specialty suppliers and are sized according to a different convention.

The thickness of hardwood lumber is specified in quarters of an inch, measured when the wood is in a rough, unplaned state. The thinnest stock is 4/4, representing 1 in., and the thickest usually available is 16/4, representing 4 in. Most suppliers will plane and straighten the edges of their stock before selling it. Expect to pay more for this, but without a fully outfitted workshop at home, you have no sensible alternative. Of course, the finished thickness is different from the rough thickness. For example, 4/4 stock ends up being 13⁄16 in. thick.

Instead of being milled to specified dimensions, like pine, hardwoods are sold in random widths and lengths. Normally, the narrowest boards are 4 in. wide and the shortest lengths are 6 ft. long. Depending on the species, boards can range up to 12 in. or 14 in. wide and 16 ft. long. This variety in sizing means that you must calculate the yield of a given board in the context of your specific project. To help you do this, take advantage of your lumber dealer's expertise. Talk to him or her about the nature of your project and be prepared with a list of the cuttings you need.

While the popular notion has always been that pine lumber is less expensive than hardwoods, that isn't the case anymore. These days you can purchase some hardwoods, like poplar and soft maple, for the same price as pine. Hardwoods like walnut and cherry, however, are much more expensive. The cherry we used for our console table project *(see page 68)* cost nearly $5 a board foot, compared to $3 a board foot for pine.

Not only are hardwoods sized differently from softwoods, but they're also priced differently. Most suppliers sell pine boards at a price per lineal foot, based on the width of the board. Hardwoods are sold by the board foot, which is defined as a square foot of rough lumber that is 1 in. thick. When a board is thicker than 1 in., the dealer multiplies the square footage by the thickness to arrive at the sale price. An 8/4 board will therefore cost twice as much as a 4/4 board of the same size.

Working with hardwoods is quite different from working with pine. With the exception of poplar and basswood, which tend to be rather soft, you cannot drive a nail through hardwood lumber without first boring a pilot hole. And cutting and planing hardwoods requires extremely sharp tools. But the resulting edges are clean and crisp. Because of this, when a project calls for fine detail work, a hardwood is the best choice.

Some hardwoods, such as oak and ash, are known as open-grain woods. These species have alternating areas of relatively porous and dense wood. The grain patterns in these boards tend to be quite striking. When stained, the open-grain areas absorb the color readily while the harder areas are more resistant. This contrast accentuates the grain patterns, creating a dramatic effect.

Cherry, maple and birch are closed-grain woods. These woods demonstrate a more uniform texture throughout a board. They are excellent choices for projects with a formal or reserved appear-

Manufactured panels are available in a wide range of surface veneers glued to different panel cores. Here, the top panel has a cherry veneer over a solid lumber core. The middle panel has a maple veneer over a veneer core. And the bottom panel has a white oak veneer over a particleboard core.

ance. Poplar is also a closed-grain wood, but its color ranges from a creamy beige to olive green, and frequently has purple highlights thrown into the mix. Because of this unusual coloration, poplar is rarely used if a furniture piece is going to have a clear finish. This wood is best when stained or even painted.

Manufactured Panels

Hardwoods are also commonly used as outer veneers on manufactured panels. These veneers are extremely thin sheets of wood that are glued to a panel core of solid wood strips (lumber core), alternating veneer layers (veneer core) or particleboard. Such panels are usually fabricated in 4 × 8-ft. sheets, but they are available in other sizes. Their thicknesses range from ¼ in. to 1 in., and they are often used in furniture and cabinet construction, and architectural trimwork.

There are many advantages to using manufactured panels. Because of their laminated construction, they are extremely stable in all dimensions. And using them yields considerable labor and cost savings, especially when large, flat surfaces are required. Since the veneers on any given panel are usually cut sequentially from the same log, the panel should display a uniform color and grain. Matching the grain pattern of solid wood to the generally uniform grain pattern on the panels can be difficult. But careful planning can yield good matches in the most visible areas of your project.

Because solid hardwoods, like the softwoods mentioned earlier, will move with changes in humidity, this must be taken into account in projects that combine solid wood with manufactured panels. Many joinery techniques have been developed to accommodate this movement. The use of solid wood frames surrounding veneered panels in door construction is just one example.

Manufactured panels do have a couple of limitations. First, whenever a panel is used, regardless of the core, the edge must be hidden. In most cases, this is achieved by gluing a strip of solid wood to the panel edge. This process is called edge-banding. And second, the veneers on the panel surface are extremely thin, often less than 1⁄32 in. Because of this, the surface is fragile and has a tendency to split out, especially on the back side of a saw cut. Also, since the veneer is so thin, there is little margin for error when sanding the surface. Aggressive sanding can quickly work through the veneer and expose the unattractive panel core underneath.

There's no doubt that many successful furniture pieces have been constructed of softwood lumber. And manufactured panels play an important role in many high-end, extremely expensive pieces. But for the beginner, the best choice is still solid hardwood stock. It may be more difficult to work with than softwood, and in some cases it doesn't have the dimensional stability of panels. But when you're done, you'll have something that will last a lifetime and look good every time you glance its way.

Text and Photos by Neal Barrett

Abrasives

Of all the activities involved in furniture making, sanding has to be the least popular. It's always messy and irritating, and usually tedious and frustrating. This is true for the beginner and expert alike and probably is the result of nothing more sinister than bad timing. After spending so many hours in stock preparation, joinery and assembly, most of us just want to get on with the finishing so we can see the piece come alive. Unfortunately the road to a good finish passes through a lot of sanding.

The most common abrasive material is sandpaper. It's generally sold in full sheets that measure 9 × 11 in. It's also available in discs and belts of various sizes to fit both portable and stationary sanding equipment. But for the beginner, the flat sheets are all you need.

Sandpaper is manufactured with various abrasives on the surface, each with a preferred use. Garnet paper is a fast sanding variety that's best suited to working by hand. However, it's not terribly durable. If you plan to use a palm sander, you'll need a longer lasting, tougher abrasive. Aluminum oxide paper is the best choice for this application.

When it comes to the finishing stage of your project when you need to sand between finish coats, silicon carbide paper is the best bet. It holds up especially well in the finer grits.

All of these sandpapers are rated by the coarseness of their abrasive particles. You can find papers ranging from a very coarse 40-grit up to an extremely fine 1500-grit. But for general work, grits that range between 100 and 320 will do the job.

Sandpaper is also classified by the type and weight of its backing material. Both paper and cloth are used for this purpose, but paper is by far the more common. When you look at the back of a sheet of sandpaper, you will see a code describing the grit of the surface and the weight of the backing. The weights are classified from A to X, with A being the thinnest and most flexible backing. For hand sanding, A-weight paper is appropriate, and for an orbital sander, C-weight paper is best.

Basic abrasives are (clockwise from top left) garnet paper, aluminum oxide paper, silicon carbide paper, wet/dry paper, steel wool and abrasive pads.

Finer-grit papers are also available with water-resistant backings. These are usually classified as wet/dry abrasives. They're most often used to level finishes like varnish or lacquer, before the final polishing. In wet sanding, water is used as a lubricant to keep the paper from clogging.

The general sanding rule is to move from coarser to finer grits in sequence, until you reach the desired finish. For most applications, you should start sanding with 120-grit paper. This should remove any scratches or other defects, like subtle planer marks, from the surfaces of the wood. But if your stock is particularly rough with pronounced, clearly visible planer marks, then you should start with 100-grit paper. In rough conditions, you can also save a lot of tedious sanding time by first lightly hand planing the surface.

A good sequence of grits for most surface finish applications is: 120, 150, 180 and 220. But if you choose a penetrating finish, like the oil finish we used on our console table *(see page 68)*, then you should finish up your sequence with 320-grit paper.

To use these papers, simply cut them in half lengthwise and wrap them around a block of wood or cork that measures about $\frac{3}{4} \times 3 \times 4$ in. Be sure to thoroughly remove the dust from your project after each grit. This is necessary because some abrasive is always broken off the backing during the sanding process. If these larger pieces are not removed, they will continue to abrade the surface at a coarser grit while you're working with a finer one.

Machine Sanding

When you have a lot of sanding to do, particularly when the stock is very rough, an orbital palm sander is the tool of choice because it's much more aggressive and less laborious than hand sanding. When using this tool, always move it parallel to the grain of the wood. Don't apply pressure to the machine—simply let its weight do the work. Move it slowly and evenly along the length of the workpiece, overlapping each stroke by one-half the width of the sanding pad. The coarsest grit should eliminate all visible scratches and defects. The succeeding grits only need to remove the scratches left by the previous paper. As with hand sanding, be sure to remove all the dust from the workpiece between grits.

Plan of Action

It's a good idea to plan your sanding before assembly. This is particularly important where two parts join to form an inside corner. If you wait until after assembly, the sanding becomes much more difficult.

Once the piece is assembled, make sure to ease all the edges. This creates a clean, crisp appearance and greatly reduces the chance of splinters breaking off the edges. To do this, hold your sanding block at a 45° angle to the edge and sand across the edge, perpendicular to the grain. Don't sand along the edge, which would be parallel to the grain. These short crossgrain strokes give you better control of the block and make it easier to achieve a uniform edge.

Remember, whenever you sand wood some dust is created, and this dust can be harmful if inhaled. Even power sanders that have dust collection attachments are never 100% effective. Always wear a dust mask when working to protect your lungs.

Other Abrasives

While sandpaper is the furniture maker's primary abrasive, there are others that come into play on just about every project. Steel wool is one. It's inexpensive, flexible and can be used for a variety of tasks. It comes in four grades: No. 0 (the coarsest), No. 00, No. 000 and No. 0000 (the finest). The two finer grades are the ones that are used most often in furniture making. They are employed to smooth finishes between coats and to polish the final finish coat. And on pieces where you plan to use a penetrating oil finish—instead of a surface finish like varnish—you can burnish the wood before finishing with No. 000 and No. 0000 steel wool. This will polish the surface of the wood to a soft shine that will be visible even after the oil is applied.

Another abrasive option is a nylon abrasive pad, which can be used for the same jobs as steel wool. The pads come in various grades, each identified by a color. Red pads are coarse, green are medium, gray are fine and white are extra fine. While steel wool pads tend to be more aggressive, the nylon pads have a much longer life. You can even wash them out and reuse them when they become soiled.

Other abrasives include powdered pumice and rottenstone. These are ground stone materials used for the final polishing of surface finishes like varnish or lacquer. After the finish is completely cured, just sprinkle a small amount of pumice on the surface and add enough water to make a paste. Then, using a felt pad, simply rub the mixture across the surface parallel to the grain. Once a uniform satin finish is achieved, wipe off the pumice with a damp cloth and dry the surface. If you want a high-gloss finish, repeat this process using rottenstone and a new felt cloth.

Text and Photos by Neal Barrett

Basic Joinery

When building furniture, there are many ways to construct joints. The simplest are those that use mechanical fasteners, like nails and screws. While these are sometimes appropriate, they're not often used in first-class work, especially in visible areas. What's preferred is a direct joint between parts, bonded with glue.

Of course, the type of joint you need depends on a variety of factors, like the nature of the materials being joined, the function of the joint, strength and appearance requirements, what machinery and equipment are available, and your own level of skill. Whole books are devoted to this discipline—and most are far from comprehensive because the possibilities are almost endless. In light of this, joinery can certainly seem intimidating to the beginner. But it doesn't have to be. By mastering two primary joints, the edge joint and the mortise and tenon, you can build an astounding array of furniture.

Making Edge Joints

The first requirement of a good edge joint is that the two mating surfaces must fit together perfectly. The mating surfaces must be flat and square to both faces of the board. This means there are no discernible gaps. Second, the mating surfaces must be either on the edge or the surface of a board. End grain is not a candidate for edge joining because of its open cellular structure. When glue is applied to these cells, they act like straws, pulling the glue deep into the wood instead of leaving it near the surface where the bond takes place. When end grain must be joined to edge or face grain, the joint of choice is the mortise and tenon. More on this later.

1 Scribe a straight reference line on one surface, using a long straightedge.

2 Clamp this board to the side of your worktable and use a bench plane to flatten the edge. Check your progress relative to your reference line frequently. And check for square frequently with a combination square.

3 Once you're satisfied with the edge on the first board, repeat the same process on the mating board.

4 When you've flattened this edge, lay the two boards together on a flat surface and check for fit. Usually some additional work will be required to get a perfect joint.

5 When the joint is perfect, spread glue on both mating edges *(see Figure A)* and clamp the boards together until the glue sets *(see Figure B).*

Making Dowel Joints

One common problem with edge joining is that the glue often acts as a lubricant between the boards. This can cause the boards to slip when clamped, which makes it difficult to achieve a flat joint. There are three common solutions to this problem: dowels, joining plates and splines. Because the last two require some fairly specialized equipment, dowels are the best choice for the beginner. All you need for the job is a drill and a doweling jig. For standard 13/16-in.-thick stock, 1/4-in.-dia. × 1-in.-long dowels are a good choice.

1 Lay out the dowel locations every 6 in. to 8 in. along the joint.

2 Install the 1/4-in.-dia. bushing in your doweling jig and center the hole in the jig bushing over your first mark.

3 Tighten the jig in place and bore a hole in the edge *(see Figure C)*. Make sure that the hole is deep enough to allow a 1/16-in. space at each end of the dowel for excess glue.

Birch Dowel Stock

Keep in mind that birch dowel stock, in 36-in. lengths and in diameters from 1/8 in. to 1 in., is commonly available at hardware stores and lumberyards. When using this material, it's a good idea to cut a narrow groove down the length of each piece to create an escape route for excess glue. You can use the corner of a sharp chisel to scratch the side of the dowel. You also should slightly bevel both ends of the dowel with a piece of sandpaper. This bevel makes aligning the dowels in their mating holes easier. You can also buy ready-made dowels from mail-order suppliers.

FIGURE A: A simple edge joint requires only the proper fit of the parts and a thin coat of glue. Be sure to spread the glue evenly.

FIGURE B: Pull the edge joint tight with clamps. Make sure that both boards are aligned correctly, so that the surface is flat.

FIGURE C: Using small dowels helps with edge joint alignment. Bore the dowel holes with a doweling jig and an electric drill.

FIGURE D: Once the holes are bored, spread glue in the holes and along the board edges. Then gently tap the dowels into place.

FIGURE E: Position the mating board so its holes align with the dowels in the first board. Then squeeze the two boards together.

FIGURE F: Begin laying out a tenon by marking a shoulder line on all four sides of the board. Use a square and a sharp pencil.

4 Repeat the same procedure for all the holes along the joint.

5 Cut your dowels to size *(see Birch Dowel Stock).*

6 Spread a thin layer of glue in all the dowel holes and along the edges of the mating boards. Then gently tap the dowels into the holes *(see Figure D).*

7 Align the mating board so the exposed dowels meet their corresponding holes *(see Figure E)* and use clamps to pull the joint tight. Tighten the clamps slowly to allow any excess glue to escape and leave the joint clamped until the glue sets.

Making Mortise-and-Tenon Joints

As mentioned earlier, the mortise-and-tenon joint is the best way to join end grain to long grain. The tenon is the male portion of the joint that is cut on the end of one board. It's designed to fit into an identically sized slot, the mortise, in the mating board.

Beginners often avoid this joint because the skills required seem out of reach. But if you take care in layout and cutting, you can easily achieve good results. Of course, it's always a good idea to practice on some scrap wood first. The tools you'll need are a combination square, marking gauge, drill, doweling jig, backsaw and sharp chisel.

1 To lay out the joint, begin by marking the tenon shoulder line. This represents the length of the finished tenon, which is usually 1 in. to $1\frac{1}{2}$ in. long. Use a square and pencil to extend this shoulder mark to both sides and edges of the board *(see Figure F).*

2 Set up your marking gauge to scribe the tenon width on the center of the board end. For $\frac{13}{16}$-in.-thick stock the tenon is usually $\frac{3}{8}$ in. thick with $\frac{7}{32}$-in. shoul-

FIGURE G: Use a marking gauge to mark the tenon lines on the edges and the end of the board. Center the tenon on the end of the board.

FIGURE H: Clamp the board upright and use a backsaw to cut along the layout lines. Be sure to cut on the waste side of the line.

FIGURE I: Once the cheek cuts are made, lay the board flat and make the shoulder cuts. Again, stay on the waste side of the line.

FIGURE J: Use a sharp chisel to remove the saw marks on the cheeks and to cut the tenon to the finished dimension.

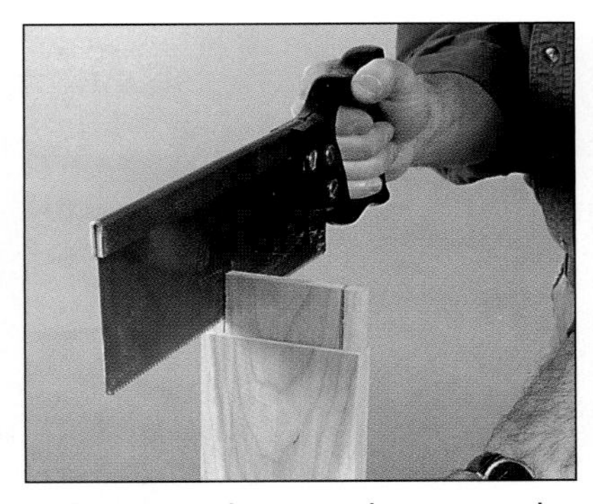

FIGURE K: Lay out the top and bottom shoulder cuts. Then clamp the board upright and make the long cuts with a backsaw.

FIGURE L: Lay the board on its edge and make the shoulder cuts on the tenon with a backsaw. Cut on the waste side of the line.

FIGURE M: Use a square and a marking gauge to lay out the mortise. Double-check that the mortise and tenon widths match.

FIGURE N: Use a doweling jig and drill to bore overlapping holes within the mortise layout lines. This removes most of the waste.

FIGURE O: Use a sharp chisel to square the ends and sides of the mortise. Work carefully to get absolutely flat surfaces.

ders on both sides. But a 5⁄16-in.-thick tenon with 1⁄4-in.-thick shoulders is also perfectly acceptable. Scribe these guide lines across the end grain and down the two edges till they meet the shoulder line *(see Figure G)*.

3 Clamp the board in place with the joint end pointing up and use a backsaw to cut along the guide lines *(see Figure H)*. Be sure your saw kerf always stays on the waste side of the line, and stop cutting when you reach the shoulder mark.

4 Clamp this board flat on your worktable and use a backsaw to cut along the waste side of the shoulder line *(see Figure I)*. When this cut is complete, the waste should fall from the side of the tenon.

5 Repeat the same process for the other side of the joint.

6 Use a sharp chisel to pare the sides of the tenon (often called the cheeks) down to the guide lines *(see Figure J)*.

7 Most tenons also have shoulder cuts on the top and bottom edges. To cut these, first lay out the guide lines using a marking gauge.

8 Cut along the length of the tenon using a backsaw *(see Figure K)*.

9 Finish by cutting along the shoulder line on both edges *(see Figure L)*. Set the tenons aside for the moment and begin working on the mortises.

10 Use the marking gauge and square to mark guide lines for the mortise in the mating board *(see Figure M)*.

11 Because cutting a mortise requires accurately removing a great deal of stock, a drill and doweling jig are your tools of choice. Just insert a bushing in the doweling jig that matches the width of your mortise.

12 Clamp the jig onto the board with the hole centered between your layout lines.

13 Slide the drill bit into the bushing and bore a series of overlapping holes until all the waste is removed *(see Figure N)*. For the best results, set the hole depth on your drill bit by attaching the collar that comes with the jig to the bit. The hole should be 1⁄16 in. deeper than the length of the tenon. This provides some space for excess glue that would otherwise keep the joint from closing completely.

14 Once all the holes are bored, square the ends and the sides of the mortise with a chisel *(see Figure O)*.

15 Test fit the joint. The tenon should be snug in the mortise, but you shouldn't have to force the parts together.

16 If the joint is too tight, carefully pare the tenon cheeks with a sharp chisel until the fit is correct. If you need to remove just a bit of stock, use sandpaper.

17 Once you're satisfied with the fit, apply glue to all the mating surfaces, and slide the pieces together. Clamp the assembly securely until the glue has dried.

Text and Photos by Neal Barrett

Finishing

For dramatic results, it's hard to beat the finishing process. With the first touch of a rag or brush, your furniture piece comes alive and the true character of its wood grain jumps out at you. Gone is the well-sanded but relatively neutral surface of the raw stock. And if you have any doubts about whether all the work was worth it, those will probably disappear, too.

The first step in the process is to pick a suitable finish. Many products are available with varying characteristics. But the one thing they all have in common is protection. All are designed to inhibit the transfer of moisture and to prevent the surface from being contaminated by dirt and stains.

The most common finishes are shellac, lacquer, varnish and oil. All are solvent- or oil-based products and therefore require some precautions. Wear protective gloves, goggles and a respirator with organic vapor cartridges when using these materials. And make sure your work area is ventilated according to the recommendations printed on the product's container.

One of the biggest differences among these four traditional finishes is how they function on wood. Shellac, lacquer and varnish are all surface finishes. This means that they do not penetrate, to any great extent, past the surface of the wood. The first coat certainly goes in the farthest, but subsequent coats merely build on the first to form a smooth, usually glossy, surface. Oil, on the other hand, is considered a penetrating finish because it reaches much deeper into the wood and leaves only a microscopic layer of finish on the surface. Subsequent coats continue to penetrate.

All these finishes have their strengths and weaknesses. Shellac, for instance, is a wonderful product. When properly applied it creates a stunning high-gloss surface which over time takes on a remarkable amber color. It's also very fast-drying, which is a great advantage over some other finishes. Unfortunately, shellac is very prone to water stains.

Lacquer is also quick-drying and is the preferred finish of many professionals. It's usually sprayed on and yields a clear, hard finish that stands up well to practically any abuse. It is, however, very flammable.

Varnishes, both the traditional types and

the newer polyurethane versions, are extremely durable and some impart a warm amber tone, not unlike shellac. Unfortunately, varnish can be difficult to apply and it takes a very long time to dry. Because of this, airborne shop dust becomes a real problem. It settles in the finish before the finish is dry. Then it has to be rubbed out before another coat is applied.

For the beginner, an oil finish is the best choice. It's easy to apply, dries fairly quickly and is not difficult to repair. It also imparts silkiness to the surface and develops a beautiful patina over time.

Preparation

Of course, proper surface preparation is one of the keys to good finishing. And the only way to get it is by sanding.

1 For the best results, you should work through a sequence of abrasives, starting at 120-grit and moving to 150-, then to 180- and 220-grit, finishing up with 320-grit. *(This process is described in more detail in* "Abrasives," *which begins on page 34.)*

2 Always dust off the piece thoroughly before moving to each new grit and ease any sharp edges by hand sanding *(see Figure A)*.

3 Once you're done sanding be sure to wipe off the entire piece with a tack cloth *(see Figure B)* or a rag that's been slightly dampened with linseed oil.

Staining

If you want to alter the natural color of a piece, you must stain or dye the wood. Oil-based stains are certainly the most common approach. These products are available at paint and hardware stores, and at home centers. They contain a pigment that is suspended in an oil and mineral spirits solution. Oil stains don't penetrate the wood deeply. They give color by embedding opaque pigment in the surface grain. You apply these stains with a rag or brush, let them sit for 10 to 15 minutes, and then wipe them off.

Aniline dyes are another approach to coloring wood. The type that is dissolved in water is the safest and easiest to use, as well as being the most colorfast. These dyes penetrate much more deeply than stains and actually change the color of the wood. It's easy to intermix colors and you can change the concentration of color simply by adding more water.

Once you've chosen your coloring method, be sure to test it out on some scrap pieces of stock before turning to your project. Taking the time to get it right is much easier than removing a stain or dye that you don't like.

FIGURE A: Ease all the edges on your project with sandpaper and a sanding block. Work across the grain for best results.

FIGURE B: When you're done sanding the entire piece, brush off all the dust. Then wipe the entire surface with a tack cloth.

FIGURE C: Apply an oil finish with a lint-free rag. Rub the oil into the wood and let it dry for 1 hour. Then wipe off the excess.

FIGURE D: Use No. 0000 steel wool to rub the finish between coats. Use a tack cloth to remove all the dust before recoating.

FIGURE E: Once the last oil coat is dry, apply a light coat of paste wax to your project. Be sure to cover all the surfaces.

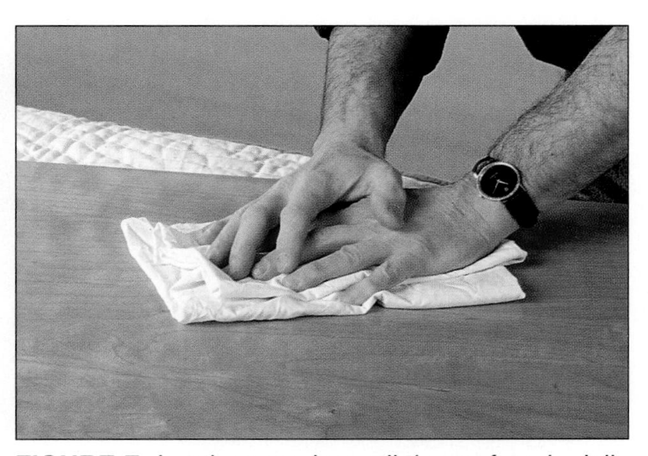

FIGURE F: Let the wax dry until the surface is dull. Then buff the entire piece to a high luster with a clean soft rag.

Applying the Finish

When you're ready to apply your oil finish, be sure to read the manufacturer's instructions carefully, and do what's recommended. While specific directions may vary, there are some general guidelines for applying an oil finish.

1 Begin by rubbing a liberal amount of oil into the surface of the wood using a lint-free rag *(see Figure C)*.

2 Allow the oil to absorb for about an hour, then wipe off the excess.

3 Let the surface dry for 24 hours, and then rub the whole piece with No. 0000 steel wool *(see Figure D)*.

4 Remove any dust, then apply another coat of oil as before. For a good finish, you should apply a minimum of three coats.

5 Once the last coat is dry and rubbed with steel wool, apply a light coat of paste wax to the entire surface *(see Figure E)*.

6 When the wax takes on a dull appearance, buff the surface to a satin sheen with a clean, dry cloth *(see Figure F)*.

Text and Photos by Neal Barrett

The combination of straight lines, flat planes and gentle curves that is the signature of the Adirondack chair is familiar even to those who have never had the good fortune to visit the Adirondack Mountains of New York state. While many variations on the Adirondack chair can be found, the essential spirit of the chair has not changed much from its earliest form as outdoor seating for modest cabins, rustic mountain hotels and elegant "great camps," as luxurious mountain retreats were known in the late 19th and early 20th centuries.

The flat back and the severely sloping seat might lead you to expect some discomfort in a chair like this. However, these seats are extremely comfortable, and they will provide you with hours of pleasant summer sitting on a porch, lawn, poolside or at the edge of a mountain lake. And they are as durable as they are comfortable. The sides of this chair also function as the rear legs and are the real foundation of the chair.

We chose white cedar lumber as the mate-

rial for these chairs. Cedar is naturally resistant to rot and insect damage. Depending on your location, you might find white or red cedar at the lumberyard. Either species is fine for this project. In either case, be fussy when going through the lumber pile, and select boards that are as close to being free of knots as you can get.

The joinery for the chair is very simple. All the joints are screwed together with galvanized deck screws. The screws are set into counterbored holes and covered with cedar plugs that are glued in place. For additional strength, especially in endgrain joints, we used polyurethane glue in addition to the screws. This glue is an excellent choice for outdoor construction because it is waterproof, and you don't have to rush during assembly. In addition to these favorable characteristics, it cures in the presence of moisture. Cedar, like most softwoods, is dried to about 14 percent moisture content, compared to about 7 percent for kiln-dried hardwood.

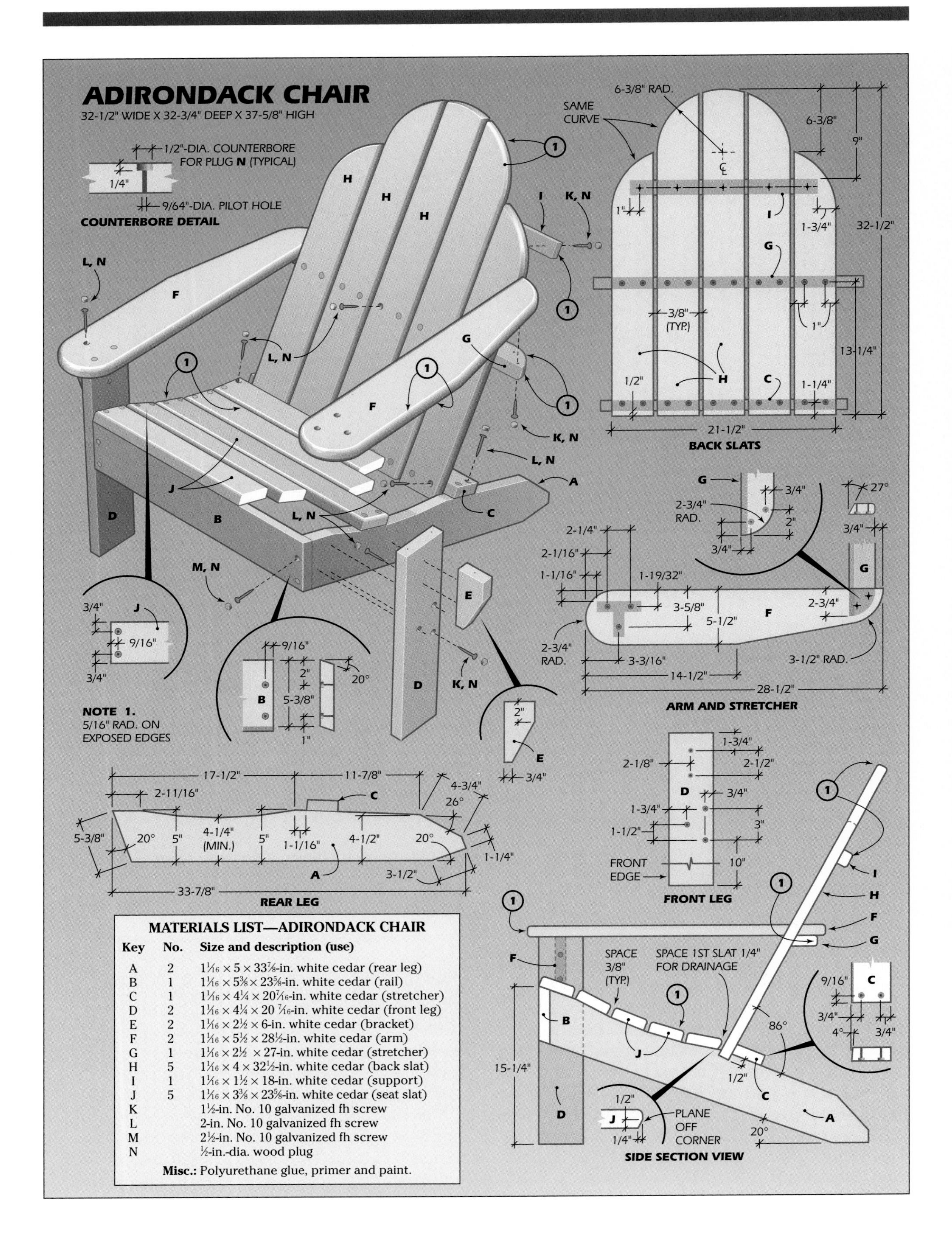

MATERIALS LIST—ADIRONDACK CHAIR

Key	No.	Size and description (use)
A	2	1 1/16 × 5 × 33 7/8-in. white cedar (rear leg)
B	1	1 1/16 × 5 3/8 × 23 5/8-in. white cedar (rail)
C	1	1 1/16 × 4 1/4 × 20 7/16-in. white cedar (stretcher)
D	2	1 1/16 × 4 1/4 × 20 7/16-in. white cedar (front leg)
E	2	1 1/16 × 2 1/2 × 6-in. white cedar (bracket)
F	2	1 1/16 × 5 1/2 × 28 1/2-in. white cedar (arm)
G	1	1 1/16 × 2 1/2 × 27-in. white cedar (stretcher)
H	5	1 1/16 × 4 × 32 1/2-in. white cedar (back slat)
I	1	1 1/16 × 1 1/2 × 18-in. white cedar (support)
J	5	1 1/16 × 3 3/8 × 23 5/8-in. white cedar (seat slat)
K		1 1/2-in. No. 10 galvanized fh screw
L		2-in. No. 10 galvanized fh screw
M		2 1/2-in. No. 10 galvanized fh screw
N		1/2-in.-dia. wood plug

Misc.: Polyurethane glue, primer and paint.

FIGURE A: Begin by making a plywood or cardboard pattern of the chair sides, and trace it onto the workpiece.

FIGURE B: Clamp the workpiece to the bench, and saw the chair sides to shape using a sabre saw held to the waste side of the line.

FIGURE C: Cut the beveled edge on the front seat support rail. Use a featherboard to prevent the workpiece from kicking back.

Making the Sides, Rails, & Arms

1 Begin by making a pattern for the chair sides/rear legs, Part A, *(see Rear Leg detail).* Trace the shape on the pattern material, and cut it out *(see Pattern Making on next page).*

2 Take the completed pattern and trace it on the side/rear leg blanks *(see Figure A).*

3 Make the angled cuts on the ends of the blanks using a sliding miter saw, table saw or circular saw.

4 Next, cut the workpiece to shape using a sabre saw *(see Figure B).* Cut to the waste side of the line and then work down to the line using a block plane and sandpaper. The finished piece should be well shaped with smooth edges that are free of saw marks.

5 Proceed now to making the front rail, Part B. Rip it to width, but make it slightly oversize (*see plan detail for finished dimension, including bevels*).

6 Cut the beveled edges on it using the table saw *(see Figure C).* Use a featherboard firmly clamped to the saw table to ensure that the

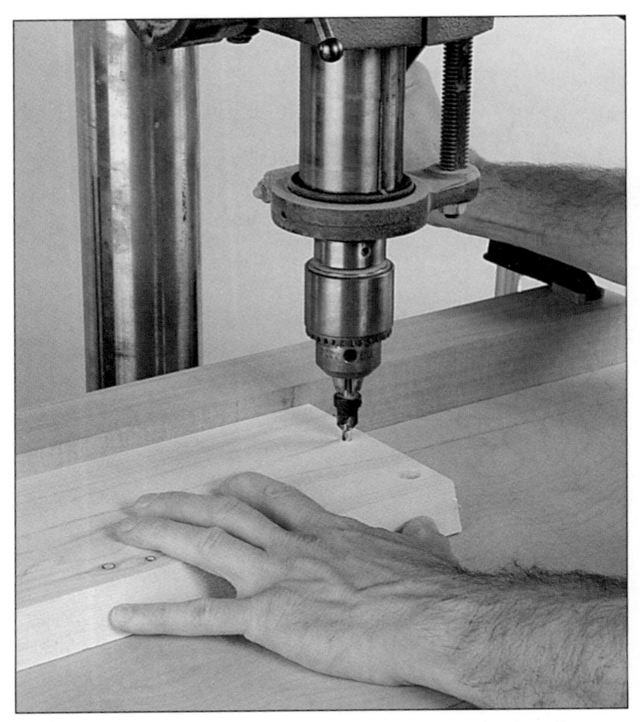

FIGURE D: Bore and counterbore pilot holes for joining chair parts using a combination tapered drill and countersink.

FIGURE E: In two stages, apply polyurethane glue to the end of the leg. Allow some glue to be absorbed, and then apply more.

workpiece moves firmly along the fence, and also to ensure that it doesn't kick back. And always use a pushstick at the end of the cut to keep your hands a safe distance from the saw blade.

7 Bore and counterbore pilot holes in the front rail for fastening it to the sides. Limit the counterbored portion of the hole to about ¼ in. deep. The most efficient tool for this is a combination drill and countersink bit chucked in a drill press *(see Figure D)*, but the holes can be made accurately with a

Pattern Making

You can use heavy cardboard or ¼-in.-thick plywood when making a pattern. Though plywood is obviously more difficult to cut than cardboard, the advantage of using it is that it is easy to make fine adjustments to the shape using sandpaper and a block plane. With cardboard, once the pattern is cut, it's difficult to adjust.

FIGURE F: Clamp the chair sides and rail in a vise, and drive the deck screws that fasten together the rail and chair sides.

FIGURE G: Spread glue on the joint between the front leg and the chair sides. Hold the parts with a clamp, and drive the screws.

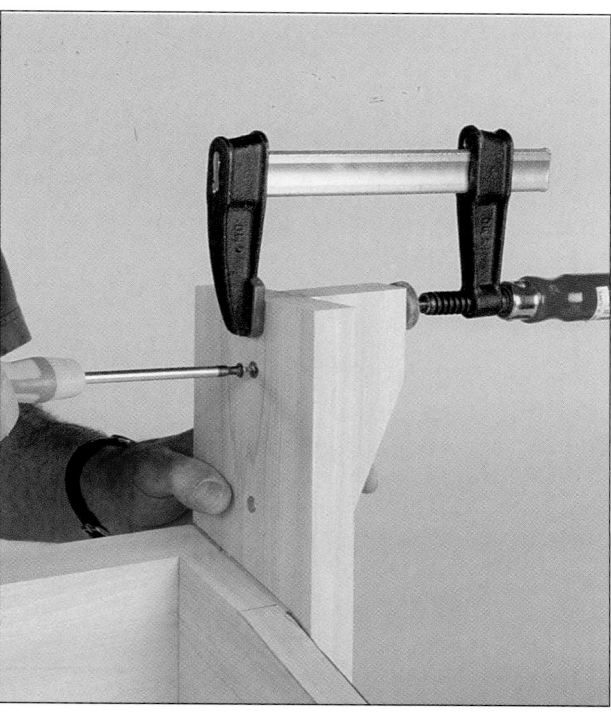

FIGURE H: Cut the arm brackets to size, and hold them in place with clamps while you attach them to the front leg with screws.

FIGURE I: Once the arms are cut to shape and their edges are rounded off, they can be attached to the front legs.

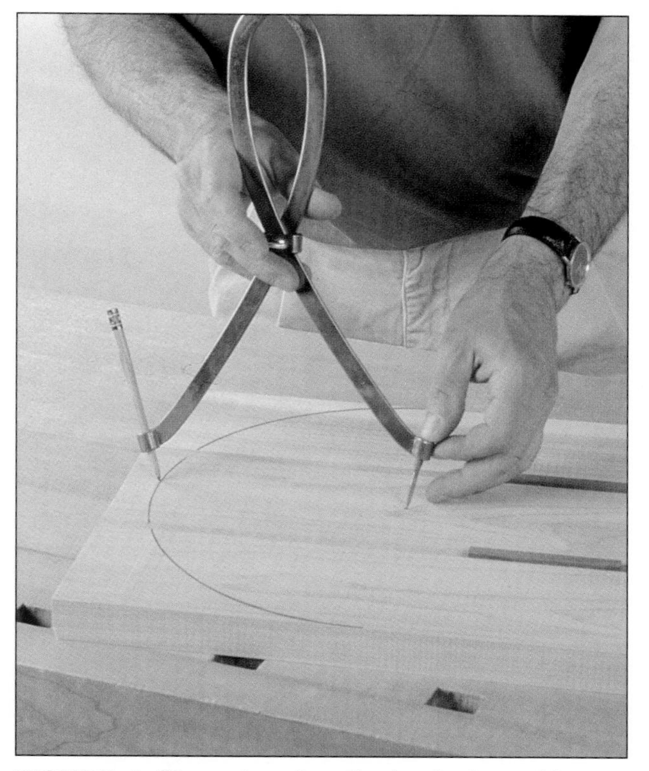

FIGURE J: Clamp together the back slats with spacers between them, and use a large compass to mark their curve.

portable drill, or a drill and drill stand.

FIGURE K: Fasten the first back slat. Place spacers between it and the next slat. Clamp them, and screw them in place.

8 Next, clamp one of the chair sides/rear legs in a workbench vise with its front end pointing up. Place the front rail over the side, and bore pilot holes into the endgrain of the side. Note: Driving screws into endgrain is generally not considered to be the best method of fastening. In this case, however, there isn't much stress on the joint, and by combining the mechanical fastening of a screw with the glue bond, we can achieve a good joint for this application.

9 Apply polyurethane glue to the end of the chair side *(see Figure E)*. There is a strong tendency for the endgrain to absorb liquid, so the best technique is to spread some glue on the piece, then wait a minute or two and reapply a bit more.

10 Position the front rail over the side, and drive the screws to fasten the two pieces *(see Figure F)*. Repeat the procedure for the opposite side.

11 Cut the back stretcher, Part C, to size.

12 Then rip the angle on its front edge as shown in the plan *(see detail on Side Section View)*.

13 Bore and counterbore the pilot holes, and apply some glue to the joints. Then fasten the stretcher to the chair sides.

14 Rip and crosscut the front legs, Part D, to size.

15 Bore the pilot holes in the legs.

16 Apply glue to the joint surfaces, and use clamps to temporarily hold the legs to the chair side assembly while you drive the screws to fasten the legs *(see Figure G)*.

17 Use a sabre saw to cut the arm brackets, Part E, and remove any saw marks with a pass from a block plane.

18 Apply glue to the brackets, clamp them

FIGURE L: After all the back slats are installed, clamp the back support in place, and attach it to the back slats with screws.

to the front legs and drill pilot holes for the screws *(see Arm and Stretcher detail for bracket position)*. Drive the screws through each leg and into a bracket *(see Figure H)*.

19 Transfer the arm profile, Part F, to the arm blanks, and cut the arms to shape using a sabre saw. Again, stay to the waste side of the line, and then refine the shape after the arm is cut.

20 Remove the sharp corner from each arm's edge, using a router and a 5⁄16-in.-rad. rounding-over bit.

21 Cut the arm stretcher, Part G, so it is slightly oversize.

22 Use the table saw to rip the angle on its front edge.

23 Trace the radius profile on either end of the stretcher *(see Arm and Stretcher detail)*, and then use the sabre saw to cut the shape.

24 Use the router and rounding-over bit to round over the edges of the piece.

25 Fasten the stretcher to the underside of the arms with screws and glue. Check that the arms are square to the stretcher before fastening. Note: Since the screws on the bottom of the stretcher will not be visible, nor directly exposed to moisture, they do not need plugs. Simply countersink the screwheads slightly below the wood surface.

26 Temporarily position the arm assembly over the chair base. Cut a scrap stick to support the back of the assembly. Then bore pilot holes through the arms and into the endgrain of the leg and arm bracket.

27 Remove the arms and apply glue to the joint. Position the parts and screw them together *(see Figure I)*.

Making The Back

1 Rip the stock for the back slats, Part H, to width, but leave the workpieces overlength. They will be cut to finished length later.

2 Clamp the three center slats together with a 3⁄8-in.-thick spacer between each.

3 Use a large compass to mark the curved profile across them *(see Figure J)*.

4 Cut the curve with a sabre saw. Mark the curve on the two outer slats, and cut them to shape.

5 Use a 5⁄16-in.-rad. rounding-over bit in the router to cut a curved edge on the front and back of each slat.

FIGURE M: Use a 1⁄2-in.-dia. plug cutter chucked in a drill press to cut out the plugs that will be glued over the screwheads.

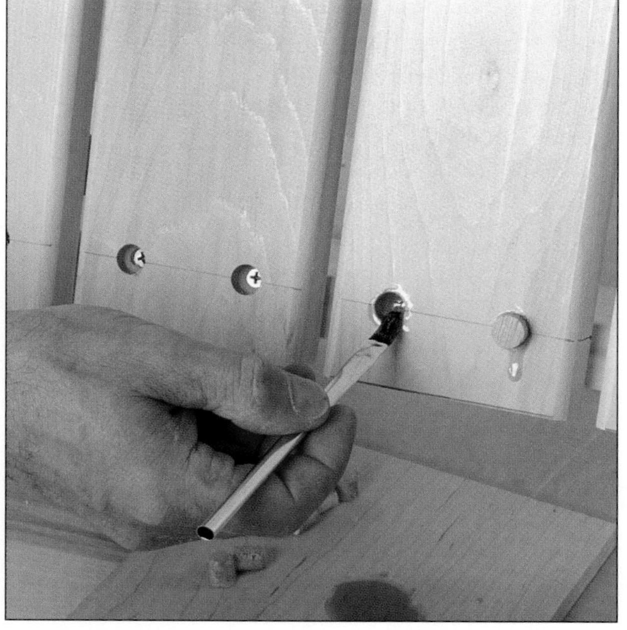

FIGURE N: Use a small disposable brush to apply a liberal glob of polyurethane glue on top of each screwhead.

6 Then crosscut them to finished length.

7 Lay out and bore the pilot holes in the slats for fastening them to the chair base *(see Back Slats detail).*

8 Hold the first slat in position on the chair, and fasten it to the stretchers with screws.

9 Clamp the second slat to the first with ⅜-in. spacers between them, and screw that slat in place *(see Figure K).*

10 Proceed across the chair back driving four screws through the front of each slat into each stretcher.

11 Cut and install the upper back support stretcher, Part I, to the back side of the slats. Use clamps to hold the part in place while you drive in the screws *(see Figure L).*

FIGURE O: Align each wood plug so that its grain matches the direction of the workpiece. Then push it into place.

Making Plugs & Seat Slats

1 Cut all the plugs, Part N, you need for the chair project at this point using a plug cutter in a drill press *(see Figure M).*

2 Use a small brush to spread a bit of polyurethane glue in each existing screw-hole, and install the plugs *(see Figures N and O).* Align the grain of the plugs with the surrounding wood to make them less visible. Note: You must install the plugs in the back slats before installing the seat slats because the screwholes will be inaccessible after the seat slats are installed.

3 Plug all the existing holes in this fashion and let the glue dry.

4 Saw each plug nearly flush *(see Figure P).*

5 Pare off the remaining material using a chisel.

6 Cut the seat slats, Part J, to size.

7 Bore and counterbore pilot holes in them *(see plan detail for hole locations).*

FIGURE P: Place a veneer shim under a fine-cutting pull-stroke saw, and cut the plug nearly flush with the surface.

FIGURE Q: Plane a bevel on the back bottom edge of the first seat slat. This creates a space that allows water to drain.

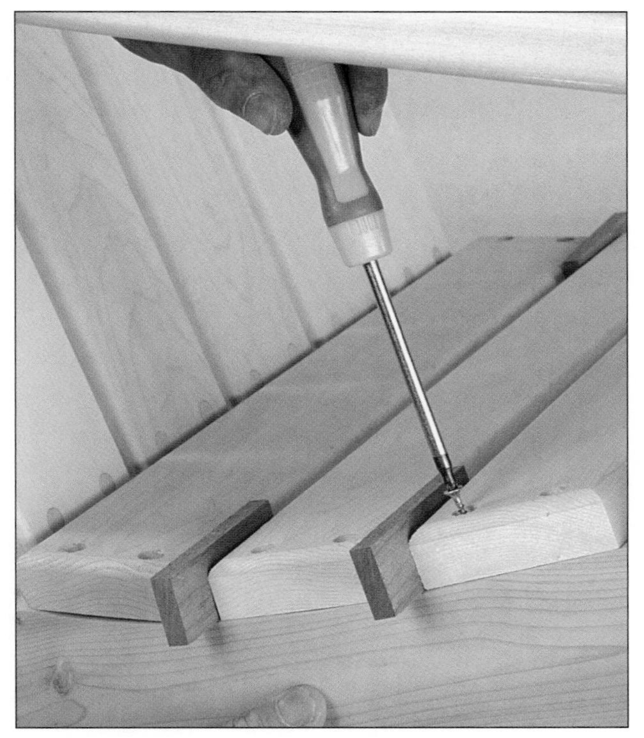

FIGURE R: Fasten the seat slats with screws. Again, note the use of 3⁄8-in.-thick blocks to ensure uniform spacing between slats.

8 Round over their top edges as you did with the other chair components.

9 Plane a bevel on the back bottom edge of the first seat slat to create a drainage space where the slat meets the chair back *(see Figure Q).*

10 Install the seat slats by screwing them to the chair sides with 3⁄8-in. spacers between them *(see Figure R).*

11 Plug these screwholes as you did the others, and proceed to finish the chair.

Finishing the Chair

1 Sand the chair with 120-grit sandpaper to remove rough spots and machine marks from the face of the lumber. Dust off the chair thoroughly.

2 Apply an oil-based primer and two coats of oil-based gloss exterior paint to all the chair's surfaces, including the leg bottoms. Make certain to follow manufacturer's instructions for drying times between coats.

Text and Photos by Neal Barrett
Illustration by Eugene Thompson

Red Cedar Table & Chair Set

Outdoor dining is one of life's great pleasures and a pastime with almost universal appeal. From crowded sidewalk cafes in Paris to the quiet backyard barbecues of Middle America, people love to gather, celebrate and dine outside. While preparing the meal is a great part of the fun, the real pleasure comes in the leisurely dining, and that's where our project shines.

Our table and chairs provide a great spot for outdoor dining. Whether you indulge in elegant or simple fare, this set provides a wonderful place to enjoy it. The table combines an ample serving surface with an intimate seating arrangement, and the chairs are extremely sturdy, yet lightweight. For added comfort, the chairs are dimensioned to work with outdoor seat cushions. These

FIGURE A: Make the bending form template with a plunge router on a trammel arm. Cut an arc in a sheet of MDF.

FIGURE B: The remaining pieces of the form are trimmed to size using the template, router and flush-trimming bit.

FIGURE C: Use ¾-in.-thick spacers between the bending form pieces. Clamp the form pieces together and fasten alignment strips.

FIGURE D: Resaw the ⅛-in.-thick apron laminate strips on the band saw. Use a pushstick at the end of the cut.

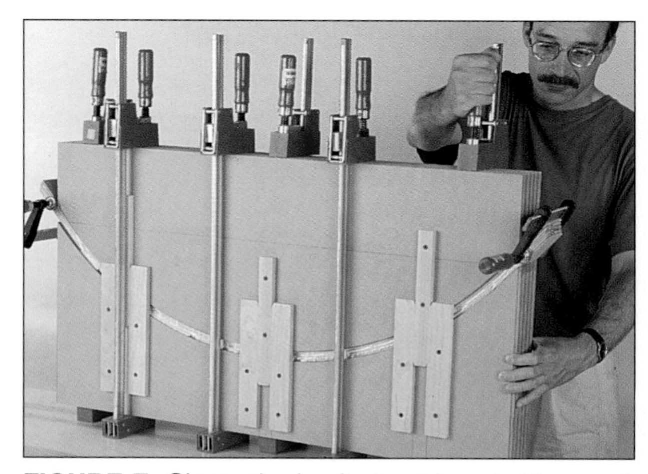

FIGURE E: Clamp the laminate strips at either end to keep them from shifting. Apply pressure with equally spaced clamps.

FIGURE F: Make a cradle. Then crosscut the apron blank to finished length. The apron length and cradle arc length are equal.

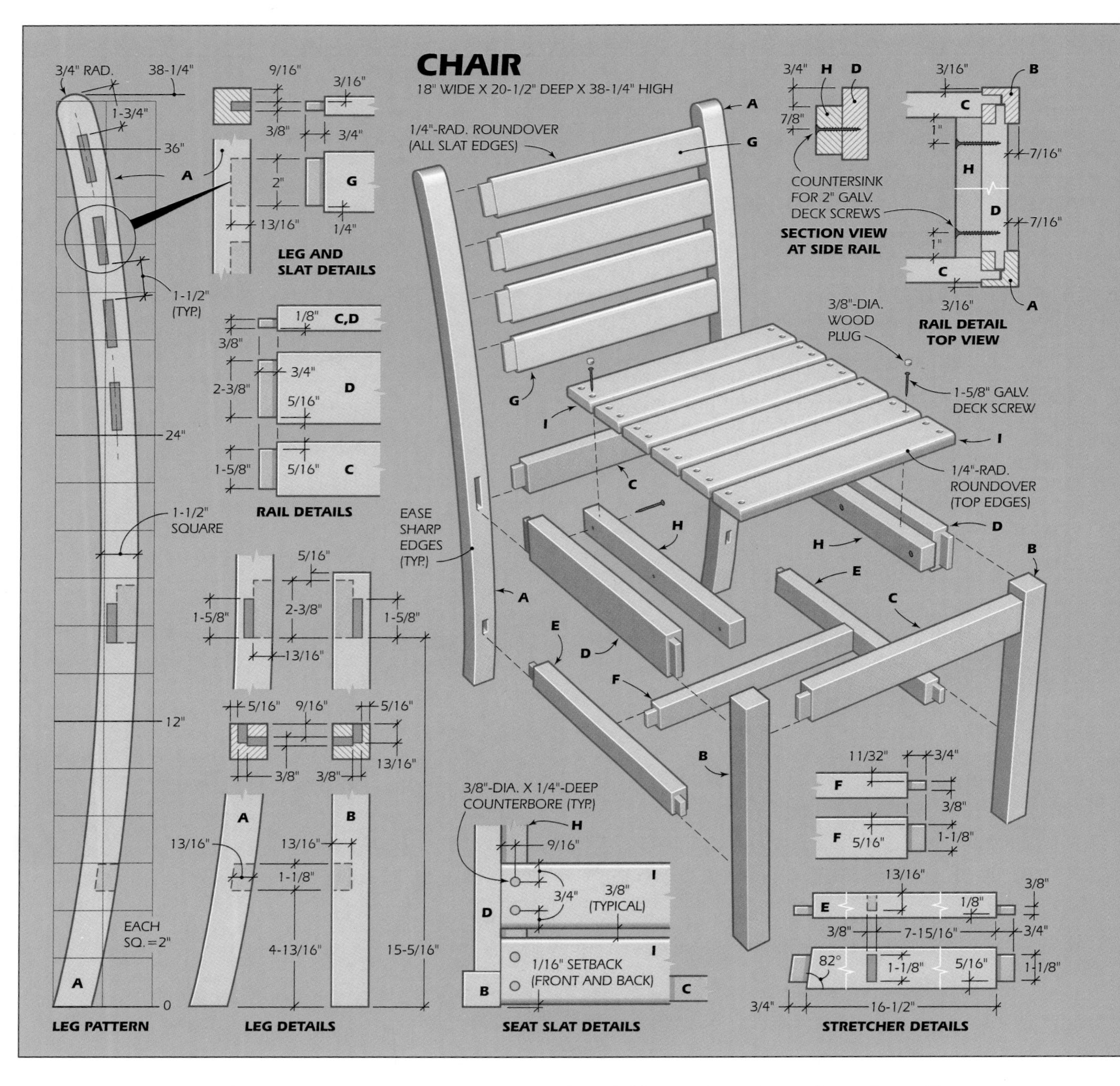

are available from Gardeners Eden, P.O. Box 7307, San Francisco, CA 94120; 800-822-9600. The table and chairs are built from red cedar, a wood known for its resistance to rot and insect damage. The pieces need little maintenance. Note that their slat construction allows water to drain off them. If left untreated, the chairs will weather to a pleasant shade of gray, but their surface will become rough. To maintain their appearance, apply an exterior sealer every year or two.

Building the Laminating Form

A laminating form *(see Apron Bending Form in the Technical Illustration)* is necessary for creating the aprons, Part J. We chose MDF (medium-density fiberboard) for the form because it is inexpensive.

1 First, make the trammel base for the router *(see Figure A for the shape of the base).*

2 Install a ¾-in.-dia. straight bit in the router, and bore a ⅜-in.-dia. hole through the tram-

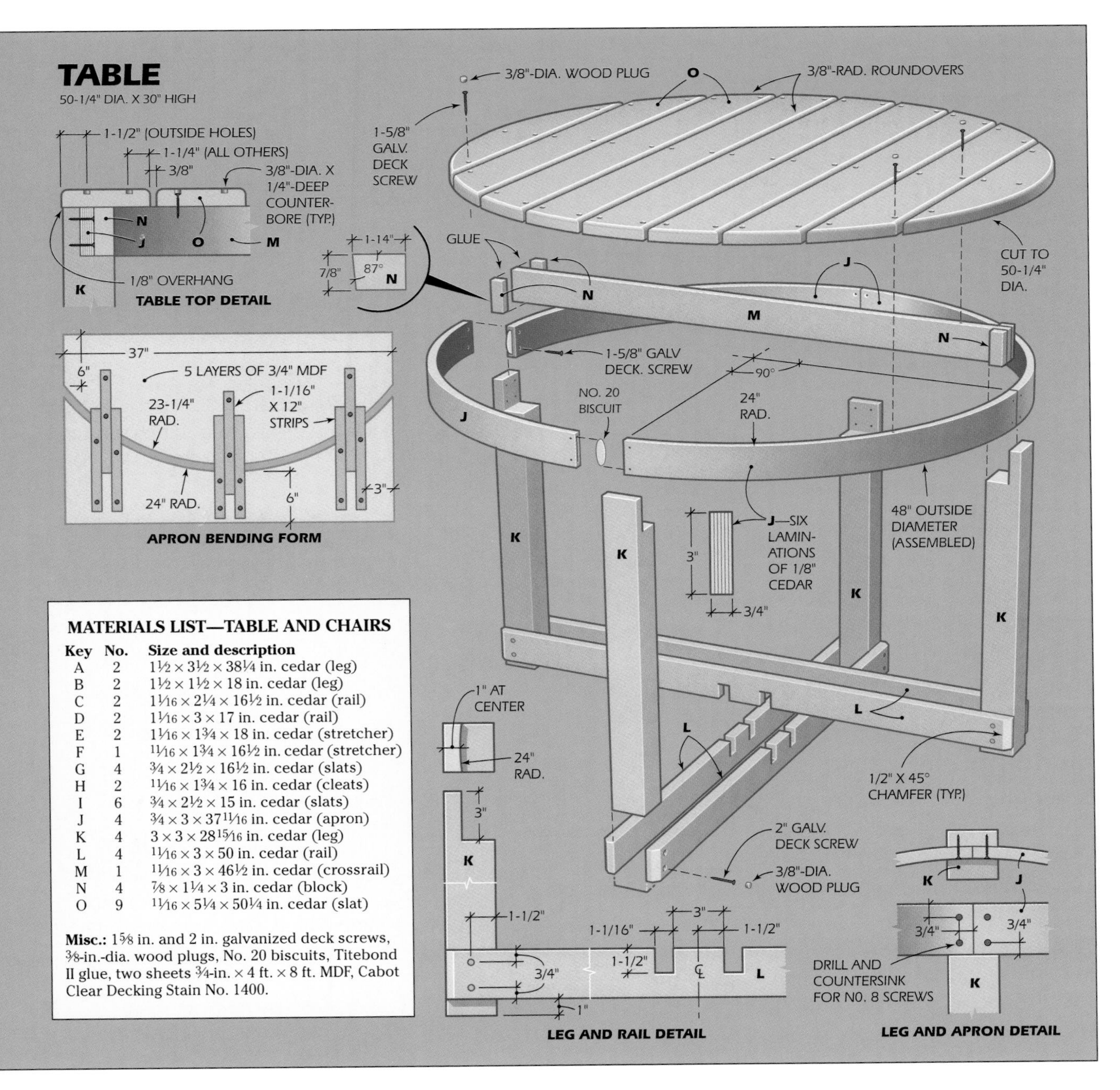

MATERIALS LIST—TABLE AND CHAIRS

Key	No.	Size and description
A	2	1 1/2 × 3 1/2 × 38 1/4 in. cedar (leg)
B	2	1 1/2 × 1 1/2 × 18 in. cedar (leg)
C	2	1 1/16 × 2 1/4 × 16 1/2 in. cedar (rail)
D	2	1 1/16 × 3 × 17 in. cedar (rail)
E	2	1 1/16 × 1 3/4 × 18 in. cedar (stretcher)
F	1	11/16 × 1 3/4 × 16 1/2 in. cedar (stretcher)
G	4	3/4 × 2 1/2 × 16 1/2 in. cedar (slats)
H	2	11/16 × 1 3/4 × 16 in. cedar (cleats)
I	6	3/4 × 2 1/2 × 15 in. cedar (slats)
J	4	3/4 × 3 × 37 11/16 in. cedar (apron)
K	4	3 × 3 × 28 15/16 in. cedar (leg)
L	4	11/16 × 3 × 50 in. cedar (rail)
M	1	11/16 × 3 × 46 1/2 in. cedar (crossrail)
N	4	7/8 × 1 1/4 × 3 in. cedar (block)
O	9	11/16 × 5 1/4 × 50 1/4 in. cedar (slat)

Misc.: 1 5/8 in. and 2 in. galvanized deck screws, 3/8-in.-dia. wood plugs, No. 20 biscuits, Titebond II glue, two sheets 3/4-in. × 4 ft. × 8 ft. MDF, Cabot Clear Decking Stain No. 1400.

mel so that the hole's center is 24 in. from the outside of the router bit.

3 Use a short length of 3/8-in. dowel to pin the trammel to a large piece of MDF.

4 Now, make three passes with the router to cut an arc through the stock *(see Figure A)*. Temporarily leave a section of the panel connected at each end of the arc.

5 Make a set of alignment marks across the arc, and use the router to cut the panel into two sections.

6 Use the two sections as templates. Cut slightly oversize blanks from the remaining MDF panel stock.

7 Screw a template to each blank, and use the router with a flush-trimming bit to cut the blanks to finished radius *(see Figure B)*. Each routed piece becomes the pattern.

8 To prevent glue from sticking to the form, apply a coat of varnish to it. Then wax it after the varnish dries.

9 Next, place 3/4-in.-thick blocks between the

bending forms, and temporarily clamp the forms together.

10 Fasten alignment strips to the surfaces of the forms *(see Figure C)*.

> ## Buying Cedar Stock
>
> We used air-dried, clear red cedar for our project. While normally we use kiln-dried stock for woodworking, we couldn't locate kiln-dried material in the sizes we needed. Besides, using kiln-dried lumber is not that important for outdoor furniture because these pieces are subjected to wide variations in humidity. To stabilize the air-dried stock, we brought it into the shop and stacked it neatly in a dry space out of direct sunlight, with evenly spaced strips of wood between each board. This is known as stickering.

Building The Table

1 Set up the band saw with a tall rip fence and a ½-in.-wide, four-tooth-per-inch blade.

2 Rip ⅛-in.-thick, 48-in.-long cedar strips *(see Buying Cedar Stock)* for the aprons, Part J, *(see Figure D)*.

3 Spread glue on the strips, and place the six strips stacked in the form. Clamp the form together *(see Figure E)*.

4 When all the apron blanks have been glued up, plane a square, straight edge on each blank.

5 Then rip the apron blanks to finished dimension.

6 Next, make a plywood cradle with a radius that matches the apron's finished outside length.

7 Clamp the cradle to a long auxiliary fence attached to the table saw's miter gauge, and place one apron blank in the cradle. The first cut removes one rough end from the apron *(see Figure F)*. Turn the apron around, and crosscut the apron to finished length.

FIGURE G: Transfer the cradle to a bench, and use it to hold the apron section in place while cutting the biscuit slots.

FIGURE H: Glue and clamp the apron sections together using a strap clamp. Check its diameter at several points.

FIGURE I: Glue and clamp together the half-lapped rail assembly. Check that the parts are square to one another.

FIGURE J: Position the apron so each of its joints is centered on a leg. Use four screws at each joint to attach the apron to the legs.

FIGURE K: Space the boards equally, and screw them to the crossrail. Draw the outline of the top on the boards.

8 Cut the remaining apron blanks to finished length.

9 Use the cradle again to hold each apron as you cut the biscuit slot in both ends *(see Figure G)*.

10 Apply glue to the apron ends, the biscuit slots and the biscuits, then assemble the apron. Use a band clamp to apply clamping pressure *(see Figure H)*. Check the apron diameter for distortion, and adjust it if necessary.

11 Rip, joint and crosscut the leg stock, Part K, to finished dimension.

12 To cut the curved notch in the leg, first make a 90° cut and then use a sharp chisel to pare the curve.

13 Rip, crosscut and notch the table rails, Part L, and chamfer the edges.

14 Spread glue on the notches, and clamp the pieces together *(see Figure I)*.

15 Position a table leg between a pair of rails, and counterbore the screw holes.

16 Fasten the legs and rails with galvanized deck screws.

17 Center a leg over each apron joint. Countersink the screw holes, and drive screws into each leg *(see Figure J)*.

18 Now cut the crossrail, Part M, and blocks, Part N, to size.

19 Place the crossrail into the leg assembly, and glue the blocks in place at its ends.

20 Rip and crosscut the top slats, Part O,to size.

21 Use a rounding-over bit in the router to ease the slat edges.

22 Clamp the center slat in position, bore its pilot holes and fasten it to the apron.

23 Fasten the remaining slats to the center rail spaced ⅜ in. from each other.

24 Mark out the top's diameter (*see Figure K*).

25 Cut it to shape with a sabre saw.

26 Sand the slat ends smooth, then use the router and rounding-over bit to ease their edges.

27 Use a plug cutter in your drill press to make the plugs to cover the screwholes.

28 Glue the plugs over the screwheads, and use a chisel to pare the plugs smooth.

29 Sand the table smooth with 120-grit sandpaper.

Making the Chairs

1 Make a thin plywood template for the rear leg, Part A *(see Leg Pattern in the Technical Illustration).*

2 Rip and crosscut the rear leg blanks.

3 Trace around the pattern onto the leg *(see Figure L).*

4 Cut the outside of the leg to shape, and smooth its outline with a block plane *(see Figure M).*

5 Cut its inside surface to shape, and smooth it with a spokeshave.

6 Rip, crosscut and plane the other chair components, Parts B through I, to final dimension.

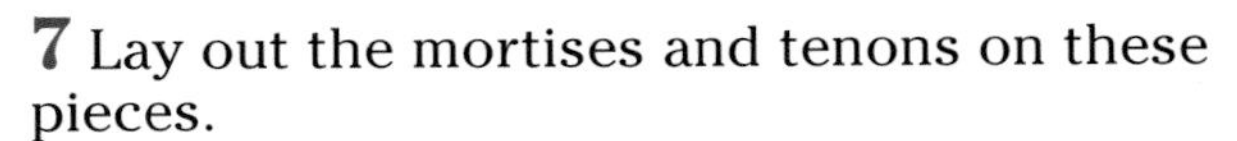

7 Lay out the mortises and tenons on these pieces.

8 Cut the mortises (except those on the inside surface of the rear legs) most easily with a router and a spiral up-cutting bit *(see Figure N)*. This will require that you cut the ends of the mortises square with a chisel.

9 This process will not work on the inside surfaces of the rear leg because the router fence does not have a straight edge to bear against. Cut these mortises by laying the leg against a fence on a drill press table. Bore a series of overlapping holes *(see Figure O).* Then cut the mortises square with a chisel.

10 Cut the tenons on the back slats, Part G, rails, Parts C and D, and stretchers, Parts E

FIGURE L: The first step in building each chair is to make a template for the rear leg, and trace it on the leg blanks.

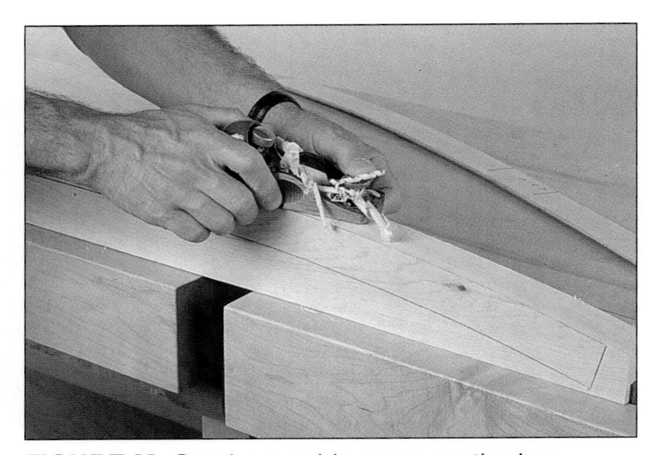

FIGURE M: Cut the outside curve on the leg. Clamp it to the bench, and smooth the curve with a block plane.

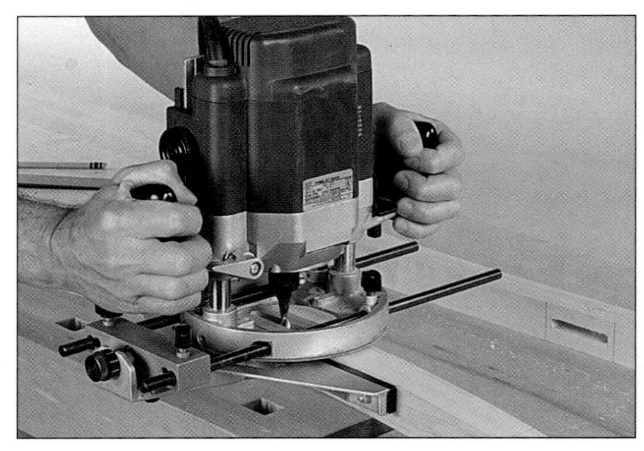

FIGURE N: Use a plunge router with its fence positioned on the leg's straight face. Cut the side rail and stretcher mortises.

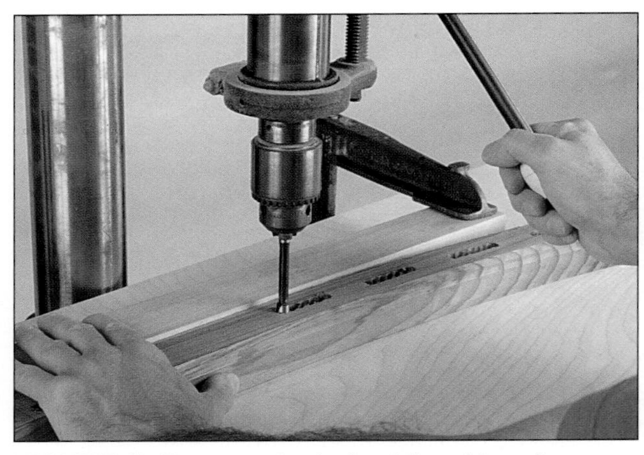

FIGURE O: Remove the bulk of the side rail mortises on the drill press. Chisel the mortise sides and ends square.

FIGURE P: Clamp a stop to the miter gauge fence. Use a dado blade to cut the tenons on the rails, stretchers and back slats.

FIGURE Q: Glue and clamp together two side stretchers with a cross stretcher. Check the assembly for square.

and F, using a dado blade installed in the table saw *(see Figure P).* On the rails and stretchers, be careful to keep track of which face of the component you are working on because the tenon is not centrally positioned on these pieces. Adjust the height of the dado blade accordingly. Also, note that the tenon that joins the side stretcher to the rear leg has an angled shoulder. Cut this by hand using a dovetail saw or backsaw.

11 Begin the final assembly by gluing and clamping together the side stretchers and the cross stretcher *(see Figure Q).* Measure diagonally from both corners of the assembly to check it for square.

12 Next, glue and clamp together the rear legs, slats and rail *(see Figure R).*

13 Glue and clamp the front legs and rail.

14 Then, glue and clamp together all the subassemblies *(see Figure S).*

15 Cut and install the cleats and the seat slats.

16 Cut and install wood plugs as you did for the table.

17 The chairs and tables were finished with a clear coat of Cabot Decking Stain No. 1400.

Text and Photos by Neal Barrett
Technical Illustration by
Eugene Thompson

FIGURE R: Clamp together the rear legs, a rear rail and four back slats. Use one clamp at each joint location.

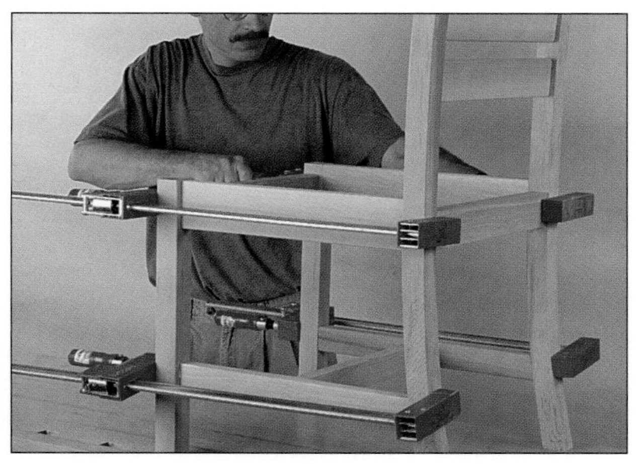

FIGURE S: Glue and clamp together the rear leg subassembly, the front legs and the stretcher subassembly.

Being equipped for warm weather fun doesn't just mean that you know where your tennis racket is or that the pool toys are inflated and afloat. For many of us, fun in the sun takes on a more civilized quality, where time is spent motionless—or nearly so—and the height of activity amounts to turning a page or lifting the nearest icy drink.

Like our more athletic friends, though, we need the right gear. And once the cooler is filled, the Ray-Bans are in place and the sunscreen's at hand, everyone knows what's needed next—a comfortable place to lay back and soak up the rays.

Our outdoor chaise lounge (or more properly, chaise longue, meaning "long chair" in French) is more than just a platform in the sun. In addition to being built to stand the abuses of the weather, it's portable, so you can follow the sun if you wish, and its three-position back suits everything from reading to cloud gazing. For creature comfort we added a long cushion, but with the chaise back fully lowered and the cushion removed, the unit converts to an attractive bench for your deck or poolside. The cushion we used is available from Gardener's Eden, P.O. Box 379907, Las Vegas, NV 89137; 800-822-9600. (Order No. 69-1611235.)

We built the chaise frame, backrest and back support out of solid redwood and assembled the components with stainless steel bolts, nuts and washers. To maintain the original color of the wood, we applied a sealant/preservative finish.

Buying Redwood Stock

Redwood is available at many lumberyards, home centers and specialty lumber dealers. The highest grade is called "clear, all heart, vertical grain." Like pine, redwood comes in nominal sizes. However, the actual thickness of the stock will be less. For example, 1x redwood stock will be about 11⁄16 in. thick and 5⁄4 redwood will be about 1 in. thick.

Building The Frame

1 Rip 5⁄4 stock *(see Buying Redwood Stock)* to width for the side rails, Part A.

2 Crosscut the rails a few inches longer than the finished dimension.

3 Make a template of the rail-end shape from 1⁄4-in. plywood and trace the shape onto each rail *(see Figure A).*

4 Use a sabre saw to cut the rail-end profiles and remove the saw marks with a spokeshave and sandpaper.

5 Lay out the mortises for joining the legs, Part B, to the rails.

6 Use a plunge router with an edge guide to make the cuts *(see Figure B).* Rout each mortise in several passes to avoid overloading the motor and bit.

7 Square the mortise ends with a sharp chisel.

8 Cut the legs and crossrails, Part C, to finished size.

9 Install a dado blade in your table saw and make the broad tenon cheek cuts on the two faces of each piece *(see Figure C).*

10 Cut the tenon shoulders in the same way.

11 Lay out the crossrail mortises on the leg inner faces.

12 Rout the mortises and square the ends with a chisel.

13 Test fit all the mortise-and-tenon joints to make sure they're snug yet go together without excess force.

14 Make a template for the leg brackets, Part D, from 1⁄4-in. plywood and trace the shape onto redwood stock. Orient the template so the grain of the stock runs diagonally.

15 Cut out the brackets as you did the rail-end profiles.

16 Mark the joining-plate positions on the legs, brackets and side rails.

17 Clamp each piece to your bench to cut the slots *(see Figure D).* Hold the plate joiner against the benchtop to ensure accurate slot registration.

18 Spread glue in the plate slots of a leg and adjoining bracket, and spread glue on the plate. Assemble the parts, clamp and repeat the process on the other legs *(see Figure E).*

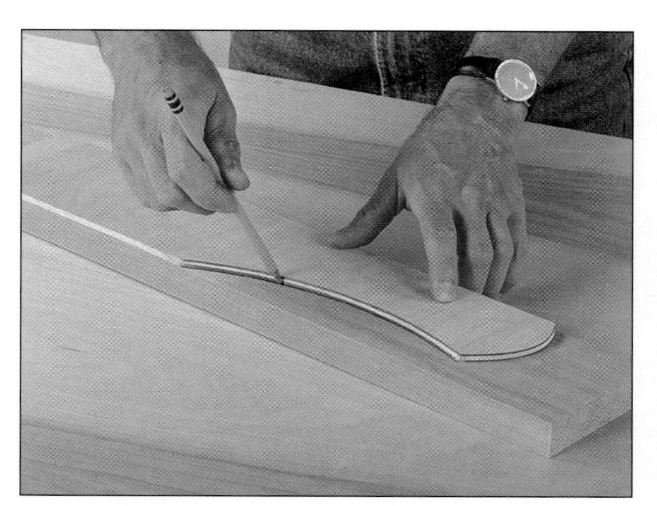

FIGURE A: Make a 1⁄4-in.-thick template for the rail-end shape. Then trace the shape onto both ends of each rail blank.

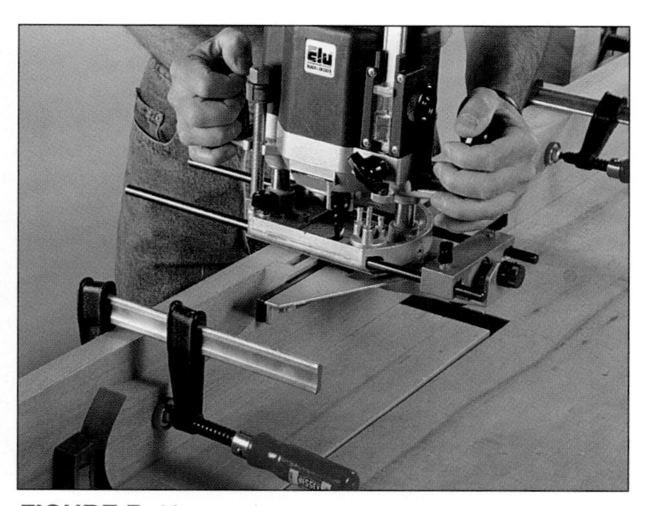

FIGURE B: Use a plunge router with edge guide to cut mortises for the legs. Clamp rails side by side to provide good router support.

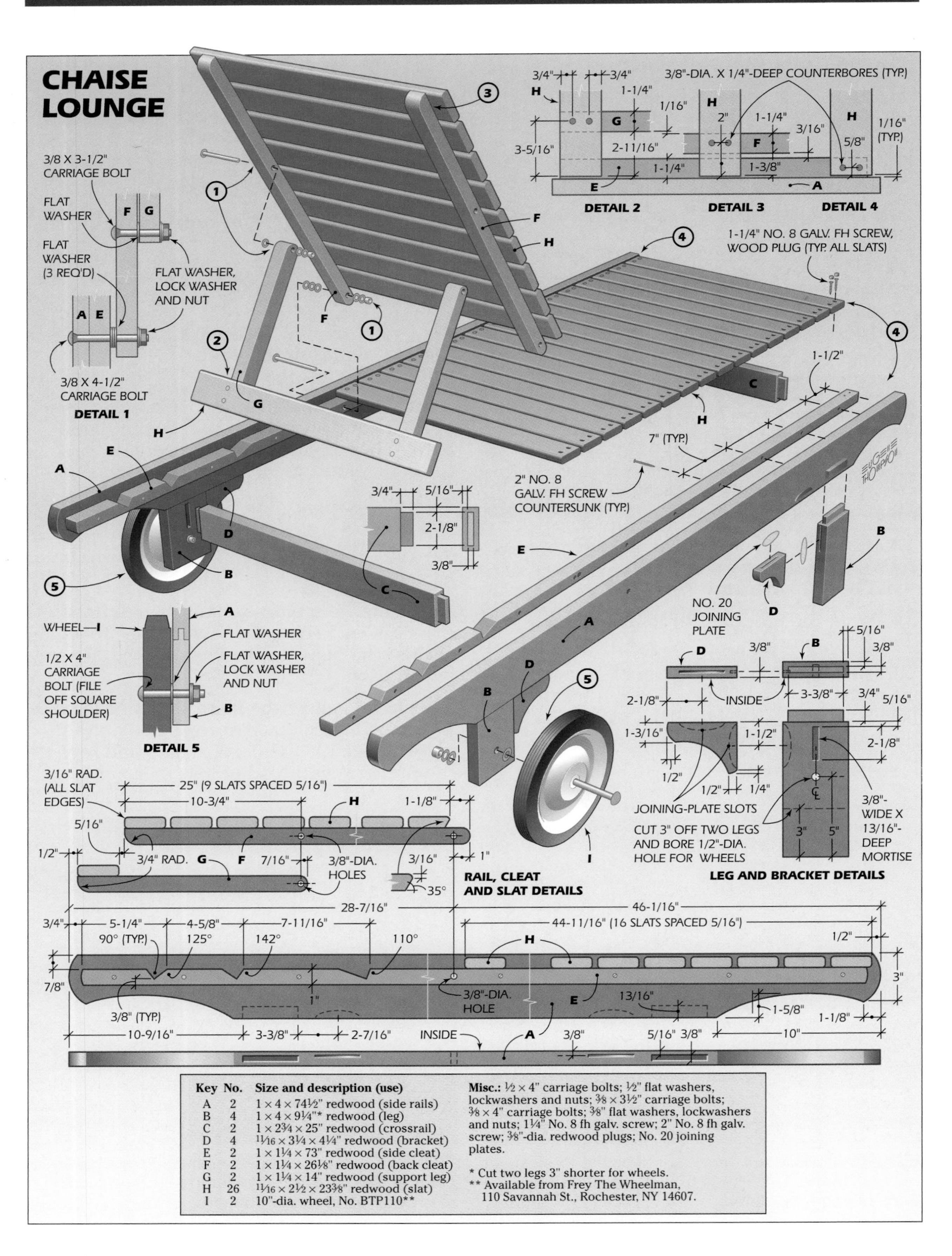

Key	No.	Size and description (use)
A	2	1 × 4 × 74½" redwood (side rails)
B	4	1 × 4 × 9¼"* redwood (leg)
C	2	1 × 2¾ × 25" redwood (crossrail)
D	4	11⁄16 × 3¼ × 4¼" redwood (bracket)
E	2	1 × 1¼ × 73" redwood (side cleat)
F	2	1 × 1¼ × 26⅛" redwood (back cleat)
G	2	1 × 1¼ × 14" redwood (support leg)
H	26	11⁄16 × 2½ × 23⅜" redwood (slat)
I	2	10"-dia. wheel, No. BTP110**

Misc.: ½ × 4" carriage bolts; ½" flat washers, lockwashers and nuts; ⅜ × 3½" carriage bolts; ⅜ × 4" carriage bolts; ⅜" flat washers, lockwashers and nuts; 1¼" No. 8 fh galv. screw; 2" No. 8 fh galv. screw; ⅜"-dia. redwood plugs; No. 20 joining plates.

* Cut two legs 3" shorter for wheels.
** Available from Frey The Wheelman, 110 Savannah St., Rochester, NY 14607.

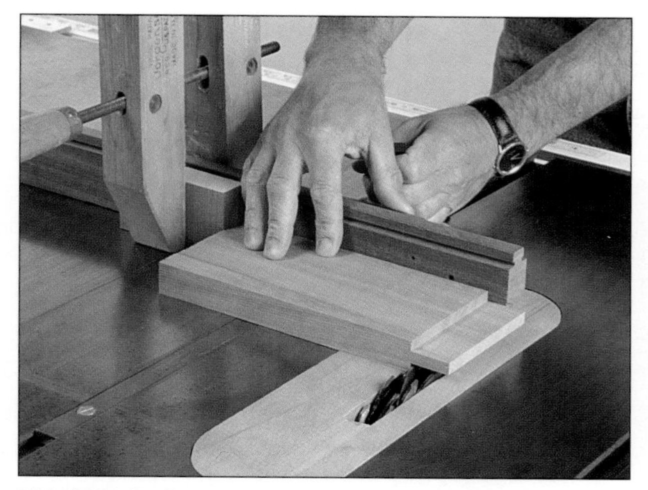

FIGURE C: Cut the leg tenons with a dado blade and table saw. A stop block clamped to the miter gauge ensures consistent tenon lengths.

FIGURE D: Cut plate joint slots for attaching the brackets to the legs and rails. Register the joiner and work against bench surface.

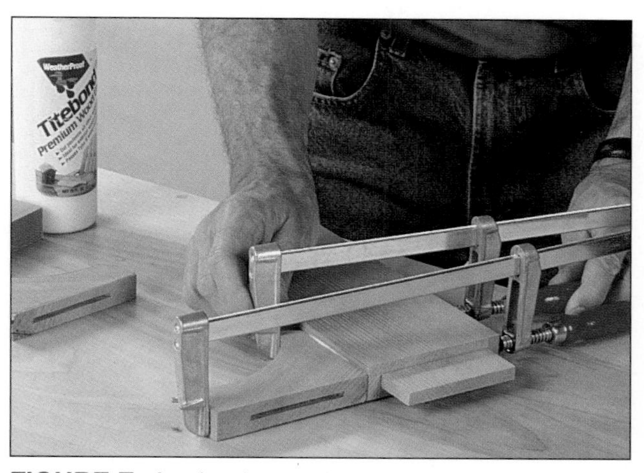

FIGURE E: Apply glue to the joining-plate slots and the plates, and join each bracket to a leg. Clamp until the glue sets.

FIGURE F: When leg/bracket assemblies are finished, join them to the rails. Apply glue to plate and mortise-and-tenon joints, then clamp.

FIGURE G: After cutting rail cleats to length and shaping backrest notches, glue and screw each cleat in place on its rail.

FIGURE H: Use carriage bolts to join the backrest cleats to the support legs. Use a washer between parts to create 1/16-in. space.

19 Prepare to join a leg/bracket subassembly to a rail by spreading glue on the mortise-and-tenon joint mating surfaces, in the two plate slots and on a joining plate. Join the leg and bracket to the rail and clamp *(see Figure F)*. Repeat the procedure for each leg.

20 Rip 5/4 stock to width for the rail cleats, Part E, and cut them to length so their ends match the rail ends.

21 Lay out the angled notches for the back-support assembly.

22 Make the cuts with a sabre saw, and sand each notch to remove the saw marks.

23 Use an exterior glue and galvanized screws to fasten the cleats to the side rails. Countersink the screwholes so the screwheads are just below the wood surface *(see Figure G)*.

24 Mark the locations of the carriage bolts that fasten the chaise back to the side rails and bore the holes through the side/cleat subassembly.

25 Apply glue to the crossrail joints, assemble the base frame, and clamp until the glue sets.

Making The Back And Slats

1 Cut the backrest cleats, Part F, and support legs, Part G, to size.

2 Use a sabre saw to trim the ends of each piece to the profiles shown in the drawing.

3 Then bore the bolt holes and join the support legs and back cleats with the bolts. Use one flat washer between each leg and cleat *(see Figure H)*.

4 Cut the slats, Part H, to size.

5 Rout the long edges of each slat with a 3/16-in.-rad. rounding-over bit.

6 Bore screw pilot holes in each slat and counterbore for 3/8-in.-dia. plugs to cover the screws.

7 Install the bolts to hold the back cleats to the frame sides. Use three washers between each back cleat and main side-rail cleat.

8 Clamp the cleats to the chaise sides. Then screw the first slat to the back-support legs *(see Figure I)*.

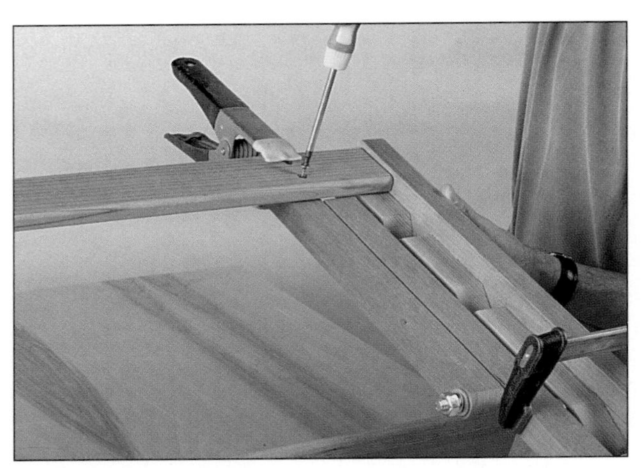

FIGURE I: Bolt the backrest and support leg assemblies to the rails. Align and clamp the parts while attaching the first slat.

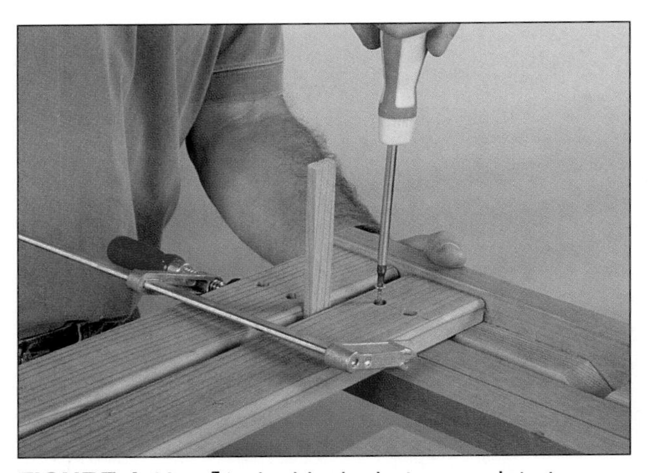

FIGURE J: Use 5/16-in. blocks between slats to create uniform spaces and screw slats in place. Keep ends of slats 1/16 in. from rails.

FIGURE K: Where the backrest joins the fixed seat portion, leave a 1-in. space so the backrest has room to pivot to its highest position.

FIGURE L: Use a small brush to spread glue on plugs. Align plug grain with grain of slat. Trim plugs with a fine saw and a sharp chisel.

FIGURE M: To provide clearance for the wheels, cut 3 in. from each rear leg with a sabre saw. Smooth the sawn edges with sandpaper.

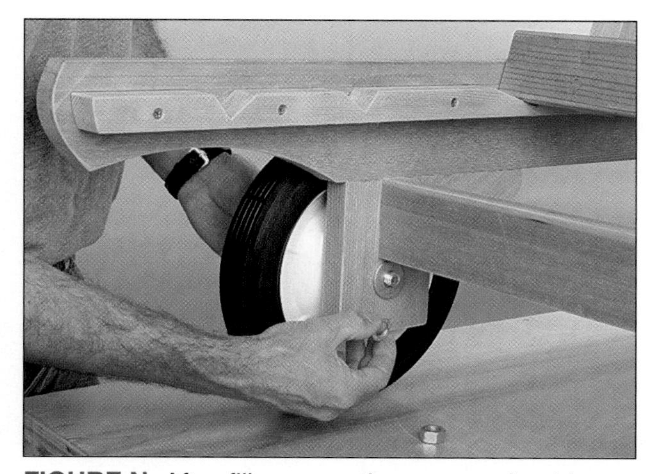

FIGURE N: After filing away the square shoulder on 1/2-in. carriage bolts, use the bolts as axles to support the wheels.

9 Use 5/16-in. blocks to space the slats and screw each slat in place *(see Figure J).*

10 Bevel the last slat on the back to provide clearance for the back to move to the highest position. Leave a 1-in. space between the beveled slat and the first slat on the chaise seat *(see Figure K).*

11 Use a 3/8-in.-dia. plug cutter in a drill press to cut plugs for all the slat screwheads.

12 Glue the plugs in place so the grain of each aligns with the slats *(see Figure L).*

13 When the glue is dry, saw each plug close to the surface, and pare flush with a sharp chisel.

14 Lay out and bore the holes in the legs for mounting the wheels, Part I, as shown in the technical illustration.

15 Then cut these legs 3 in. shorter *(see Figure M).*

16 Use a file or small grinding wheel to remove the square shoulder on two 1/2 x 4-in. carriage bolts.

17 Install the wheel with the modified bolts, washers, lock washers and nuts *(see Figure N).*

18 Sand all surfaces with 120- and 220-grit sandpaper.

19 To protect the redwood, we applied a coat of Cabot's Decking Stain (No. 1400, clear). Let the finish thoroughly saturate all surfaces and allow it to dry at least 48 hours.

Text and Photos by Neal Barrett

Technical Illustration by Eugene Thompson

Console Table

This simple console table is a great beginner's project. Its straightforward lines are reminiscent of Shaker pieces built over 100 years ago. But simple doesn't have to mean unsophisticated. This solid cherry piece is well-tailored, crisply built and can fit just about anywhere: your front hall, behind a living room sofa, in an upstairs bedroom, or even in your bathroom if it's blessed with enough extra space.

But good design isn't the whole story. This piece is also easy for a beginner to build. It has only nine parts: four legs, four rails and a top. And we show you how to build it with nothing more than hand tools and a few portable power tools. Everything you need is described in "Beginner's Toolbox" on page 6.

Perhaps the best part of this design, however, is that it puts to good use information that is in the Woodworking Skills section. If you start this table now, your gratification won't be delayed much longer. You should be able to finish it up in just a few weeks of spare time—even if you just learned how to sharpen a chisel or cut a mortise-and-tenon joint.

Preparing the Cherry Stock

1 The first step in preparing the lumber *(see Buying Cherry Stock)* is to crosscut all parts to rough length, a couple of inches longer than their finished lengths *(see Technical Illustration).*

2 Then check the jointed edge of each piece for flatness and square. If some refinements are required, clamp the board to the side of your worktable and use a bench plane to true the edge *(see Figure A).*

3 Next, cut the boards to finished width using a circular saw with a rip guide *(see Figure B).* Clean up any saw marks with a bench plane.

Buying Cherry Stock

The material we used for this piece is solid cherry stock that we bought flattened on both sides and jointed on one edge. You'll have to pay more for this service, but it's worth the cost. The standard thickness for this type of hardwood is 13⁄16 in.

This simple but elegant console table is a great introductory project for beginning furniture makers.

FIGURE A: Begin the top by flattening one edge of each board, using a bench plane. Make sure the edge is planed square to the face.

FIGURE B: Cut each top board to width using a circular saw and rip guide. Make sure the rip guide follows the planed edge.

FIGURE C: Lay out the location of the alignment dowels on the board edges. Then use a doweling jig and drill to bore the holes.

FIGURE D: Cover the edges and dowel holes with glue, insert the dowels and bring the boards together with pipe clamps.

FIGURE E: When the glue is dry remove any squeeze-out, then let the assembly cure. Flatten joints if necessary with a plane.

FIGURE F: Mark the finished length on both ends of the top panel. Then make the cuts with a saw and straightedge guide.

Making the Tabletop

1 Begin work by laying the boards *(see Tabletops)* on a flat surface and choosing the most attractive grain pattern by arranging the boards in several ways.

2 Then lay out the dowel locations on all the joints and bore the dowel holes using a doweling jig and a portable drill *(see Figure C)*.

3 Next, place a drop of glue in each dowel hole and gently tap the dowel in place.

4 Then spread the glue evenly on all the mating edges and push the boards together.

5 Tighten the joints, using pipe clamps *(see Figure D)*, and check that the panel is flat before letting the glue set. If it's not, readjust

FIGURE G: Mark the chamfer around the top and plane the edges to this line. A scrap block keeps the side edge from splitting.

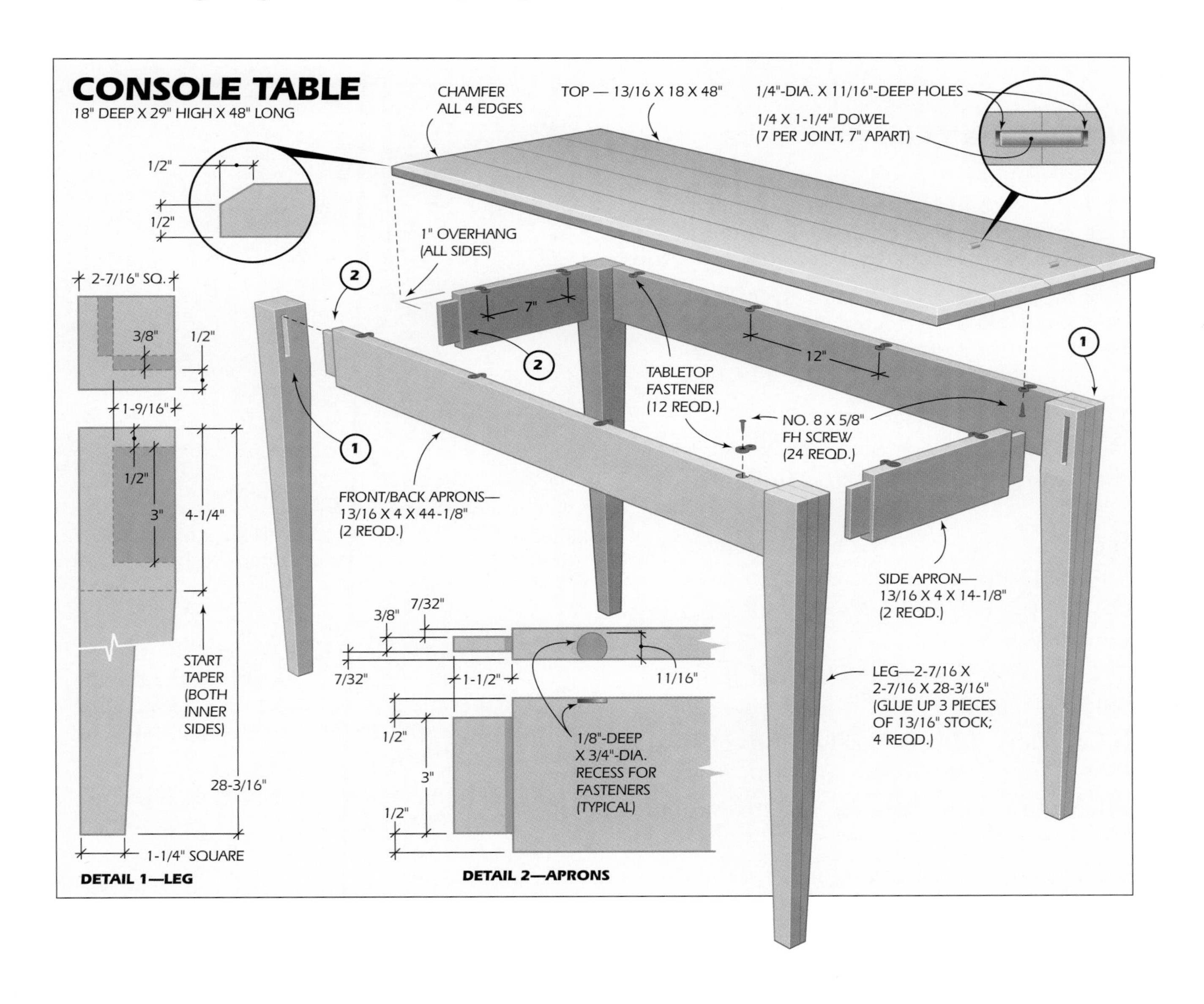

the clamps until the surface is flat.

6 After 20 minutes scrape off any excess glue from the joints and let the panel dry overnight.

7 When you remove the clamps, check the panel surface carefully. If the joints are flush, set the panel aside. If they aren't, use a bench plane to smooth the surface *(see Figure E)*. Hold the plane at a 30° angle to the wood grain and make shearing cuts.

8 Cut the panel to finished length, using a circular saw and a straightedge guide *(see Figure F)*. Make sure that both ends are square to the sides before making the cuts.

9 Then mark guidelines for the edge chamfer around the perimeter of the top and use a block plane to create these bevels *(see Figure G)*. Be sure to clamp a scrap block to each long edge to keep them from splitting when you're working on the end grain.

10 Complete the tabletop by sanding smooth both sides and all the edges. Begin with 120-grit paper and move through a sequence of 150-, 180- and 220-grits.

Tabletops

Using multiple boards helps keep a tabletop flat over time. The tabletop in this project was made from four smaller boards that were glued together. If your stock is wide enough to use only three boards, that's fine.

Making the Legs And Rails

1 Apply glue to the three boards that make up each leg and clamp them together *(see Figure H)*. Note that each leg is formed from three pieces of stock that are glued together.

2 Scrape off the excess glue after 20 minutes, and leave each leg assembly clamped for at least an hour. Don't do any further work on these pieces until the glue has cured for 24 hours.

3 Crosscut the leg stock to finished length.

4 Now crosscut the rail stock to finished length.

5 Lay out a tenon on each end, using a marking gauge *(see Figure I)*.

6 Make the tenon cheek cuts with a backsaw (see Figure J).

7 Then make the top and bottom shoulder cuts *(see Figure K)*.

8 Use a sharp chisel to refine the cuts and remove any saw marks.

9 Lay out the mortises on the corresponding legs with a marking gauge.

10 Then use a doweling jig and a portable drill to remove most of the waste *(see Figure L)*.

11 Finish up the mortise by squaring the ends and sides with a sharp chisel.

FIGURE H: Cut the leg stock to size, then apply glue to the mating surfaces. Keep the board edges flush when clamping.

12 Once the joinery is done, cut the tapers on both inside edges of each leg, as shown in the Technical Illustration. Use a circular saw and be sure to cut on the waste side of the layout lines.

13 Finish these tapers with a bench plane *(see Figure M)*, making sure to check for square as you work.

14 Before the legs and rails are assembled, it's a good idea to finish sand all the parts with the same progression of grits that was used in Making the Tabletop, step 10.

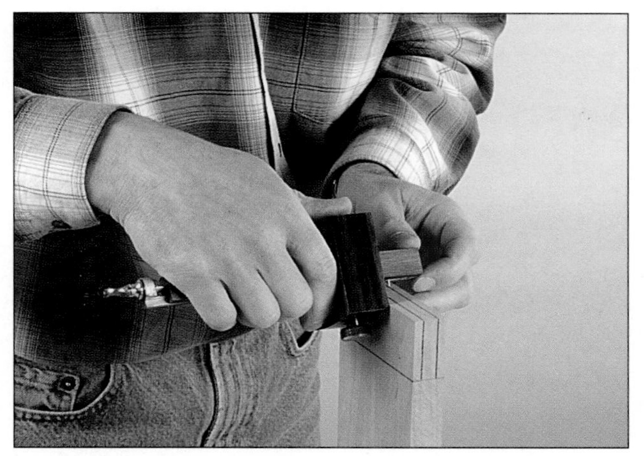

FIGURE I: Lay out the tenons on the ends of the table rails with a marking gauge. Keep the gauge base flat on the board surface.

FIGURE J: Make the cheek cuts on the tenons using a backsaw. Keep the blade kerf just to the waste side of the layout lines.

FIGURE K: Make the tenon shoulder cuts with a backsaw. Clamp a scrap block to the board to help guide the saw blade.

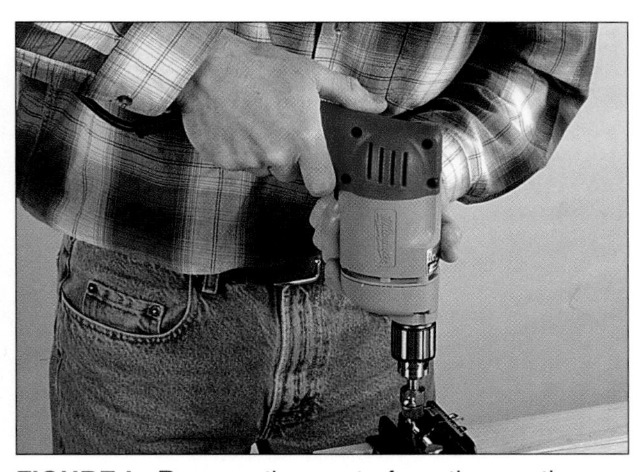

FIGURE L: Remove the waste from the mortise using a drill and doweling jig. Square the ends and walls with a sharp chisel.

FIGURE M: Rough cut the leg tapers with a circular saw. Then reduce the edges to finished thickness with a bench plane.

Assembling the Table

1 Begin by joining a long rail to a pair of legs. Spread the glue evenly on the tenons and mortises, and then clamp the pieces together.

2 Do the same with the other legs and long rail.

3 When the glue has cured on these two assemblies, join them together with the short rails. Assemble the parts on a flat surface.

4 Once the clamps are in place, compare opposite diagonal measurements to check for square *(see Figure N)*. If the assembly isn't square, readjust the clamps until it is.

5 When the base joints have cured, lay out and bore the holes in the rails for the tabletop fasteners.

6 Then turn the top upside down on a padded table and place the inverted base assembly on the underside of the top. Adjust the base so it's centered on the top.

7 Mark the location of the fastener holes.

8 Bore pilot holes and screw the base to the top *(see Figure O)*.

9 Carefully examine the parts for any scratches that occurred during assembly and sand these areas with 220-grit paper.

10 Remove the dust from the entire piece and, if you plan to apply an oil finish like we did, finish sand with 320-grit paper.

11 When you're done sanding, remove the dust again and wipe the entire piece with a tack cloth. Apply an oil finish as described in "Finishing" on page 41.

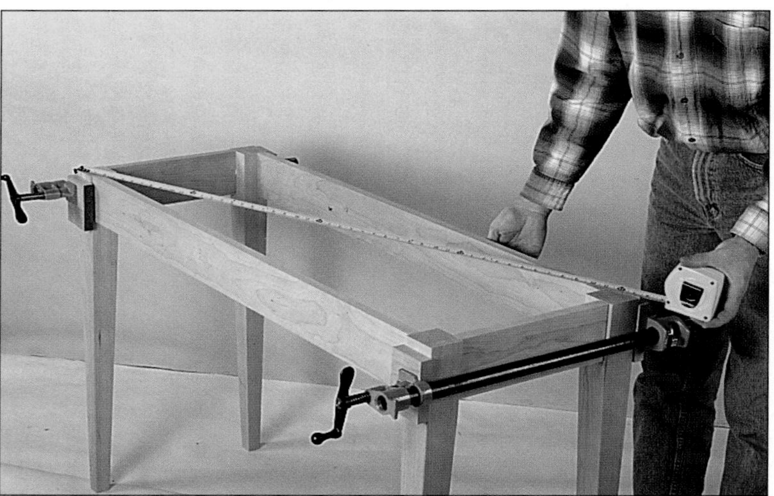

FIGURE N: Glue and clamp the legs to the rails. Then check for a square assembly by comparing diagonal measurements.

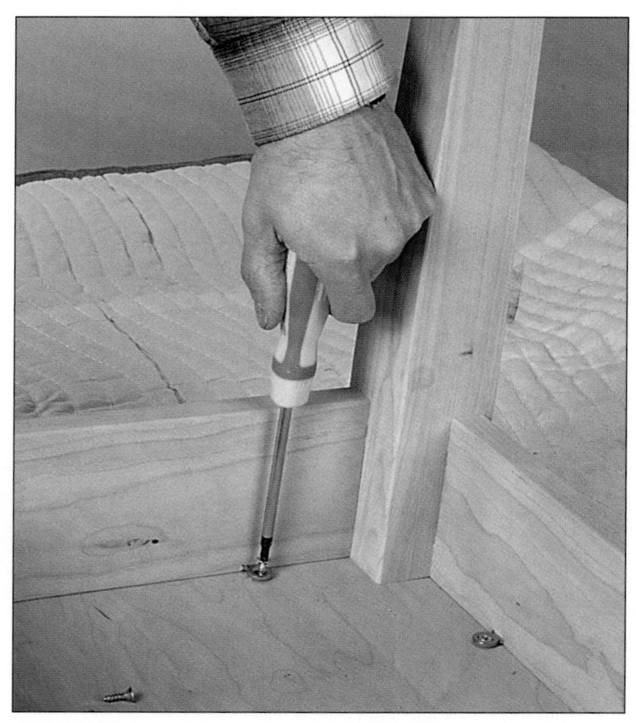

FIGURE O: Attach the tabletop fasteners to the rails. Then turn the table parts over and screw the fasteners to the underside of the top.

Text and Photos by Neal Barrett

Technical Illustration by Eugene Thompson

Building a Bookcase

High-end furniture projects are great to dream about. But unless you have a well-equipped shop and some serious woodworking experience to draw on, it can be difficult to turn the dream into a reality.

Not every piece of furniture needs to be a museum showpiece, though. Often a simple design does the job just as well and the experience gained in completing it goes a long way toward making the next project even better.

Our pine bookcase, for example, features simple construction and it's designed to be built with basic woodworking tools. Yet, the finished project is a worthy and useful addition to any room of the house. While it's meant to rest on the floor, you can convert the bookcase to a wall-mounted storage unit by leaving off the baseboard. You can secure the cabinet to the wall by screwing through the cabinet cleats into the wall studs.

We made the case out of materials available at most building-supply dealers and lumberyards, including ½ × ¾ in. parting strip, 1 × 2, 1 × 4 and 1 × 10 common pine and ¼-in.-thick lauan plywood. Assembly is quick and easy with glue and nails, and when you're done with construction you have the option of a painted or clear finish.

As for basic tools, you'll need a portable circular saw, hammer, block plane, combination square, tape measure, metal rule, two clamps, nail set and putty knife. Other supplies include glue, nails, sandpaper, wood filler and varnish or paint and shellac.

The specifications that follow will produce a bookcase with overall dimensions of 10¾ in. deep x 34 in. wide x 48 in. tall. While the depth of the case is directly tied to the 1 × 10 stock, you can vary the height, width and shelf spacing to suit your needs. Keep in mind, though, that extending the width of the cabinet may require the addition of central shelf supports.

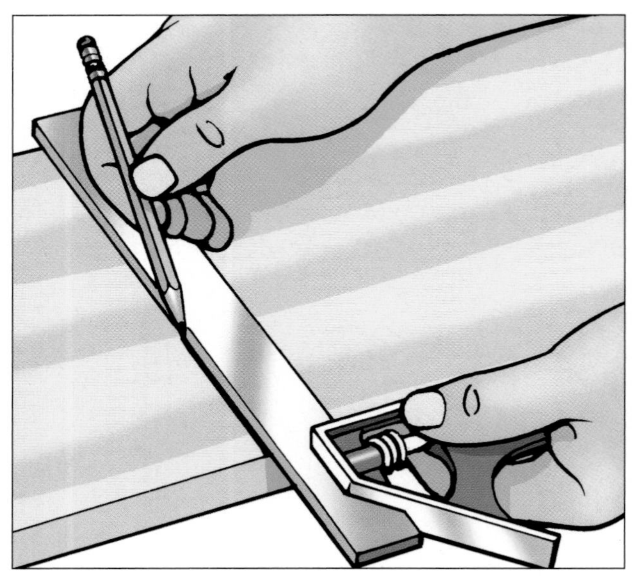

FIGURE A: Mark the lengths of the bookcase side panels on 1 × 10 lumber and use a square to lay out the crosscut lines.

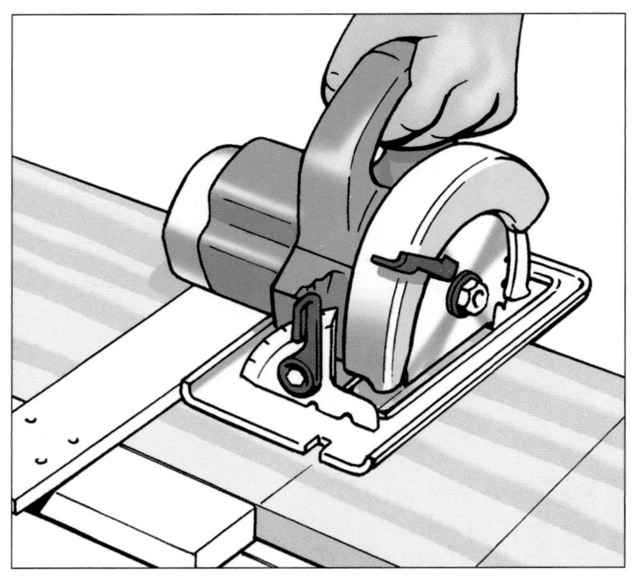

FIGURE B: With a T-guide cut to match your saw, align the end of guide with crosscut line. Tack guide to the work and then make the cut.

Cutting The Parts

1 To support your work during the cuts, use 2 × 4s spanning two saw-horses for a temporary bench and clamp your work in place. Set your circular saw cutting depth so the blade cuts about ⅛ in. into the 2 × 4s.

2 Begin construction by using a tape measure to mark the length of a side panel on 1 × 10 stock, and lay out the cut line with a square *(see Figure A)*. The side panels on our bookcase are 48 in. long.

3 Place the T-guide *(see Making a T-guide, right)* against the edge of the stock and align its trimmed end with the cut line. Tack the guide in place and use your circular saw to make the cut *(see Figure B)*.

4 After both sides are cut to length, lay out and cut the five shelves to length to suit the width of your bookcase. Our shelf length is 31 in.

5 Rip the four lower shelves to 8⅞ in. wide to allow for the thickness of the case back. Clamp each shelf to the sawhorses and tack a straight strip to the work to guide your circular saw *(see Figure C)*.

6 Next, cut the 10 shelf-support cleats from lengths of ½ × ¾-in. parting strip. Use a handsaw to cut the pieces slightly oversize.

Making a T-guide

For precise crosscuts, first make a simple, self-aligning T-guide for your circular saw.

1 Cut a piece of ½ in. plywood to 2½ × 24 in.

2 Cut a roughly 12 in. long piece of 1 × 4 pine that will serve as the crossbar of the T.

3 Center the plywood strip along the 1 × 4 and glue and screw them together, making sure the pieces are perfectly square to each other.

4 Butt the crossbar of the T-guide against the edge of a piece of scrap lumber, tack the guide in place and make a cut through the 1 × 4 with your saw base guided by the plywood strip.

5 Then, trim the 1 × 4 on the opposite side in the same way. Now, the ends of the 1 × 4 can be aligned with layout lines on the stock for precise cut positioning.

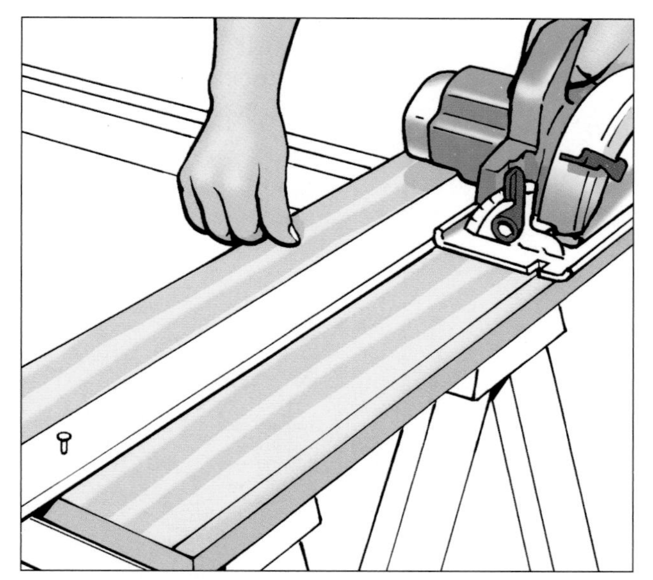

FIGURE C: Use a straight strip as a guide when ripping stock for the four lower shelves to 8⅞ in. The top shelf remains a full 1 × 10.

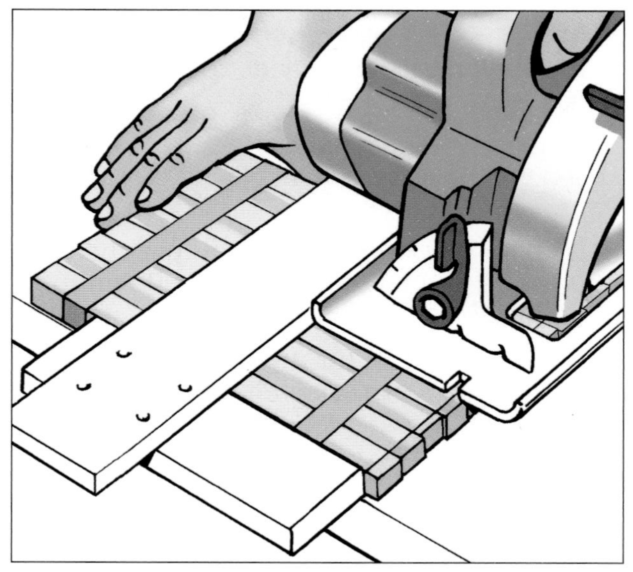

FIGURE D: After cutting shelf cleats oversize, gang them together with tape, mark the cut lines and trim with a T-guide and circular saw.

7 Then gang the pieces together with masking tape. Mark the cut lines and use your circular saw and T-guide to cut the cleats to 8⅞ in. long *(see Figure D).*

8 Because we varied the spaces between the shelves, the vertical back cleat lengths vary. Rough cut and mark the back cleats in pairs. From the top down, the lengths are 8, 9, 9½ and 11¼ in.

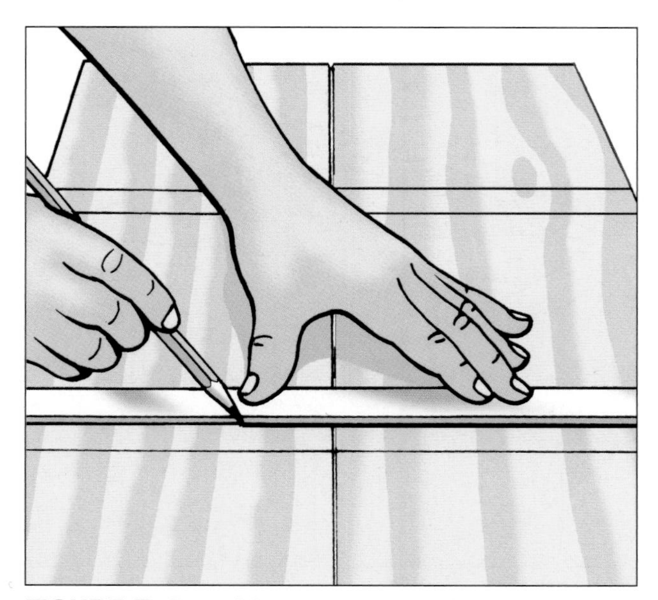

FIGURE E: Butt sides together with shelf-location marks on outer edges. Use a straightedge to extend the shelf marks across work.

9 When cutting the back cleats with the T-guide, first gang them together so all the marks on one side align.

10 After the cut, untape the cleats and reposition them so the marks on the opposite side are aligned for the next cut.

Assembly

1 First mark the shelf cleat locations. Hook your tape measure to the top edge of a side panel, extend the tape and place marks at the following dimensions: ¾, 10¼, 20¾, 31¾ and 44½ in. These marks indicate the top edges of the shelf cleats.

2 Transfer the marks to the other panel.

3 Lay both side pieces edge to edge with the marks on the outer edges and use a rule or straightedge to extend the shelf locations across both panels at once *(see Figure E).*

4 Use 2d nails and glue to attach the shelf cleats to the sides. Position the cleats so they're flush with the front edges of the sides.

5 Then, attach the vertical back cleats, leaving a ¾ in. gap at the bottom of each back cleat for a shelf *(see Figure F).* The gaps will help to keep the shelves aligned during assembly.

6 Align the back cleats with the back ends of the shelf cleats to provide the 3⁄8 in. recess for the back panel.

7 To join the sides and shelves, first lay a side panel on a few 2 × 4s placed on the floor.

8 With a helper assisting, stand the shelves in position and lay the opposite side on the shelf ends.

9 Start a pair of 6d finishing nails at each shelf location so the points just penetrate the shelves.

10 Lift the side off and apply glue to the endgrain of the shelves. Let the glue soak in for a few minutes, then apply a second coat.

11 Follow with a coat of glue on the sides and cleats.

12 Replace the panel using the nail points to align the shelves.

13 Drive the nails *(see Figure G)* and set them below the surface.

14 After the first side is attached, grasp the sides at one end while your helper grasps the opposite end and flip the assembly over.

15 Secure the remaining side and check that the case is square. If necessary, tack a diagonal brace across the back to hold it while the glue sets.

16 When the glue is dry, cut a piece of parting strip to fit between the two top cleats

FIGURE F: Nail and glue shelf cleats and back cleats to the side panels. Note that cleats are recessed 3⁄8 in. from back edge of sides.

FIGURE G: Secure one side to the shelves with glue and 6d finishing nails. Then flip the assembly over and attach the other side. Set all nails.

FIGURE H: Attach the 1 × 2 strips to the case starting with a vertical member. Then, add the horizontal pieces and the other vertical.

and under the top shelf. This piece will be set 3/8 in. in front of the top shelf rear edge to provide room for the back panel.

17 Glue and nail this long cleat to the shelf.

FIGURE I: Use a block plane to shape the chamfers on the baseboard. Rest the plane against the case to maintain a uniform angle.

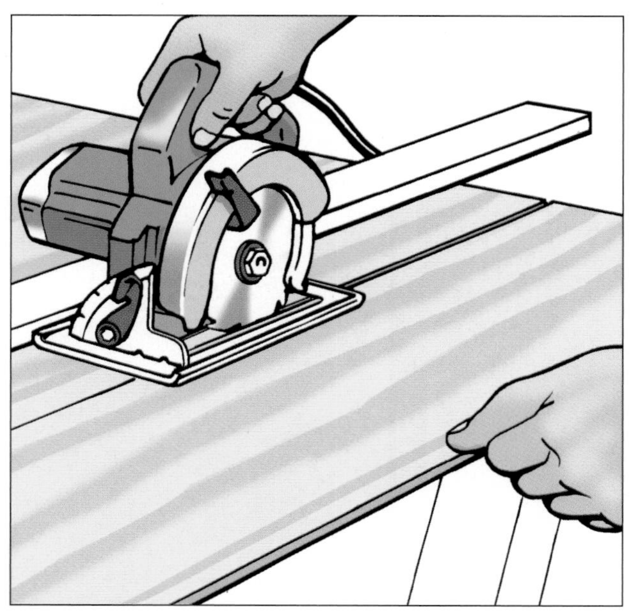

FIGURE J: Cut the 1/4-in.-thick back panel to size with a circular saw. Use a straight strip as a guide when making the cuts.

Adding The Fascia

1 To fit the 1 × 2 fascia over the case front edges, first mark the stock for crosscutting. Make the vertical pieces 48 in. long to match the sides and mark the horizontal members at 29½ in.

2 Rough cut the pieces to length and use the T-guide and circular saw to trim them squarely to exact size.

3 Apply glue to one of the vertical members and nail it to the case so its edge is flush with the side.

4 Then, add each horizontal member with glue and nails, keeping the top edges flush with the shelf tops *(see Figure H).*

5 Finally, add the remaining vertical member with nails and glue.

6 Use 1 × 4 stock for the front and two side baseboard pieces. Cut the length of each side piece to 10 in. long.

7 Glue and nail the parts in place.

8 Then, cut the front baseboard piece to 34 in. long.

9 Secure it so its ends are flush with the side pieces.

10 After the glue has dried, use a block plane to trim a chamfer around the top edge of the baseboard. Plane the side pieces first, working from front to back to avoid splitting the corners of the front piece. Then, plane the front piece to match. It helps to lean the plane against the case to maintain a uniform angle *(see Figure I).*

11 The final component is the case back. Lay out the cut lines on 1/4 in.-thick lauan plywood.

12 Tack a straight strip to the panel to serve as a guide for your circular saw and cut the stock to size *(see Figure J).*

13 Attach the back panels to the case with glue and 1 in. finishing nails.

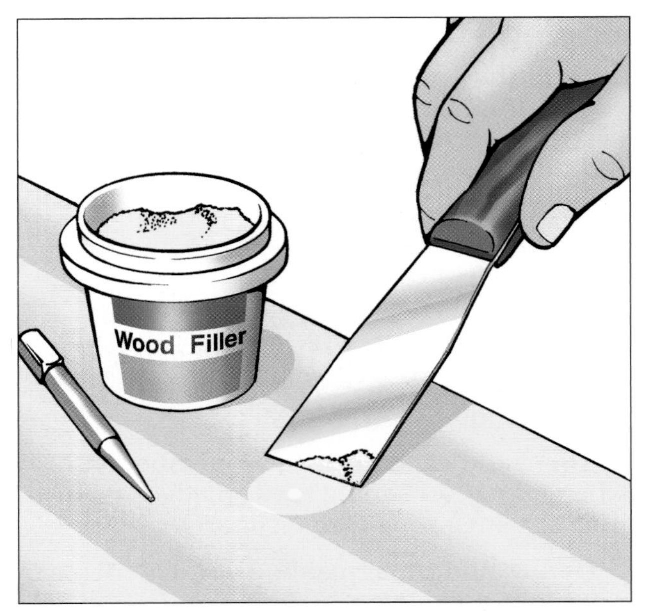

FIGURE K: Use a nail set to drive all nails below the surface. Then apply wood filler over the nail-heads with a putty knife.

FIGURE L: Apply shellac to any knots before painting. This will seal the knots and prevent sap from bleeding through.

Finishing

1 First make sure all the nails are set below the surface.

2 Use a putty knife to fill the nail holes with wood filler and let the compound dry *(see Figure K).*

3 Sand the entire bookcase with 120- followed by 220-grit sandpaper.

4 Then, use a sanding block and 220-grit paper to slightly ease all corners.

5 Thoroughly dust the case with a tack rag.

6 If you plan to paint your bookcase, first apply two coats of shellac over each knot to prevent the knots from bleeding through the final paint job *(see Figure L).*

7 Then, prime and paint the bookcase according to the paint manufacturer's instructions.

By Rosario Capotosto
Illustrations by George Retseck

Platform Bed

Beds are for sleeping, right? Well, not if you're a kid. If you're a kid, bed is the place to read, dream, plot and scheme—or just while away a rainy afternoon. It's gossip central after school is out and the intensive care unit during a bout with the flu. It's the place to return to when the day is done—and sometimes, it's just the best place to be alone.

Now that you know it's special, the job is to build a bed that's up to snuff. Our design takes care of all the basics, with a little extra just for fun. It's made of birch plywood and poplar and we've designed it to go together without a cabinet shop full of clamps. Underneath, there's ample storage, with enough space for a pile of games, books and the family cat. If you want drawers under the platform bed instead of the open compartments we show, then simply use the instructions for the dresser drawers (page 93) to help you along your way.

Text and Photos by Neal Barrett
Technical Illustration by Eugene Thompson

Building a Kid's Bedroom Suite

We've incorporated a strong but simple visual theme for our bedroom suite, the rest of which you'll find on pages 88 to 108. We've also tried to make these pieces affordable by using inexpensive poplar stock and birch plywood. But this doesn't mean that these projects won't challenge your woodworking skills. While the painted surface is a good way to hide mistakes, a first-class finish still requires topnotch stock preparation and joinery. And, if you really want to create something special, build everything out of highly figured hardwood with a clear finish.

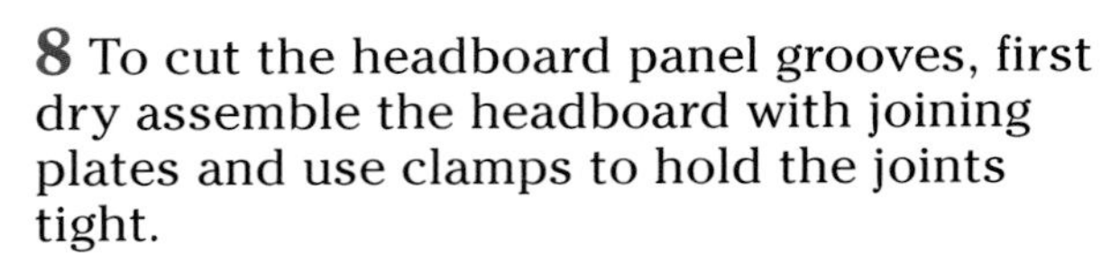

Building the Headboard

1 Cut 1½ x 3½-in. poplar to length for the headboard stiles, Part A, mullions, Part B, and top rail, Part C. To make accurate cuts with a circular saw, use a Speed Square to guide the cut *(see Figure A)*. Position the square the appropriate distance from the cutline based on your saw's base plate. Then clamp the square in place.

2 Cut the stock for the wide bottom rail, Part D, by crosscutting the available widths a few inches longer than finished dimension.

3 Apply glue to the mating edges of each piece and use clamps to pull the joints tight. Add clamps across the thickness of the assembly at each seam—at the ends especially—to help keep the boards aligned *(see Figure B)*.

4 After about 20 minutes, scrape off any excess glue, and let the glue set for at least 1 hour before removing the clamps.

5 Cut the glued-up panel to finished length using a straightedge guide clamped across the work to guide your circular saw.

6 Mark the locations of joining plate slots for the headboard joints. Note that these joints are formed by a double row of plates.

7 Using a flat table as the registration surface, hold both the joiner and workpiece against the table and cut the slots nearest one face *(see Figure C)*. Then flip each piece over to cut the remaining slots.

8 To cut the headboard panel grooves, first dry assemble the headboard with joining plates and use clamps to hold the joints tight.

9 Install a ¼-in. piloted slotting cutter in your router (Bosch cutter No. 85520, arbor No. 82811) with the pilot bearing mounted on the top of the bit arbor. Before routing the actual headboard stock, make a test groove in a 1½-in.-thick block to make sure your router is set up correctly.

10 Rout a ¼-in.-wide × ½-in.-deep groove around each panel opening and ½ in. from the stock face *(see Figure D)*.

FIGURE A: Crosscut the 1½-in.-thick poplar headboard pieces to length. A square clamped to the work helps guide the cut.

FIGURE B: Apply glue to the wide headboard rail pieces and clamp. Use clamps at the ends to help keep the faces aligned.

FIGURE C: To cut the double plate slots, use a flat surface to register the slots on one side. Then flip over work for remaining slots.

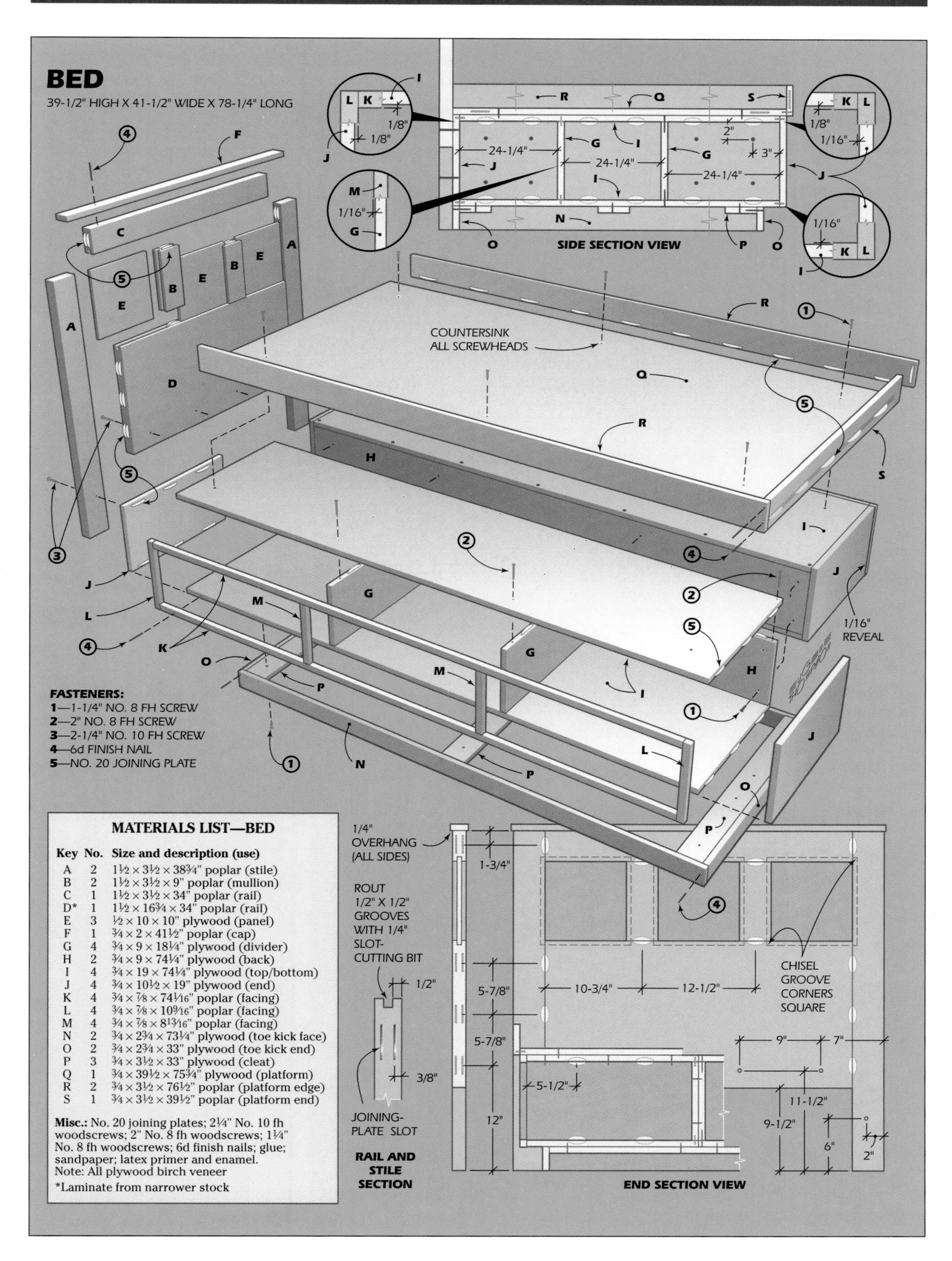

MATERIALS LIST—BED

Key	No.	Size and description (use)
A	2	1½ × 3½ × 38¾" poplar (stile)
B	2	1½ × 3½ × 9" poplar (mullion)
C	1	1½ × 3½ × 34" poplar (rail)
D*	1	1½ × 16¾ × 34" poplar (rail)
E	3	½ × 10 × 10" plywood (panel)
F	1	¾ × 2 × 41½" poplar (cap)
G	4	¾ × 9 × 18¼" plywood (divider)
H	2	¾ × 9 × 74¼" plywood (back)
I	4	¾ × 19 × 74¼" plywood (top/bottom)
J	4	¾ × 10½ × 19" plywood (end)
K	4	¾ × ⅞ × 74$^{1}/_{16}$" poplar (facing)
L	4	¾ × ⅞ × 10$^{9}/_{16}$" poplar (facing)
M	4	¾ × ⅞ × 8$^{13}/_{16}$" poplar (facing)
N	2	¾ × 2¾ × 73¼" plywood (toe kick face)
O	2	¾ × 2¾ × 33" plywood (toe kick end)
P	3	¾ × 3½ × 33" plywood (cleat)
Q	1	¾ × 39½ × 75¾" plywood (platform)
R	2	¾ × 3½ × 76½" poplar (platform edge)
S	1	¾ × 3½ × 39½" poplar (platform end)

Misc.: No. 20 joining plates; 2¼" No. 10 fh woodscrews; 2" No. 8 fh woodscrews; 1¼" No. 8 fh woodscrews; 6d finish nails; glue; sandpaper; latex primer and enamel.
Note: All plywood birch veneer
*Laminate from narrower stock

FIGURE D: Dry assemble the headboard with plates and install clamps. Rout the panel grooves with a piloted slotting bit.

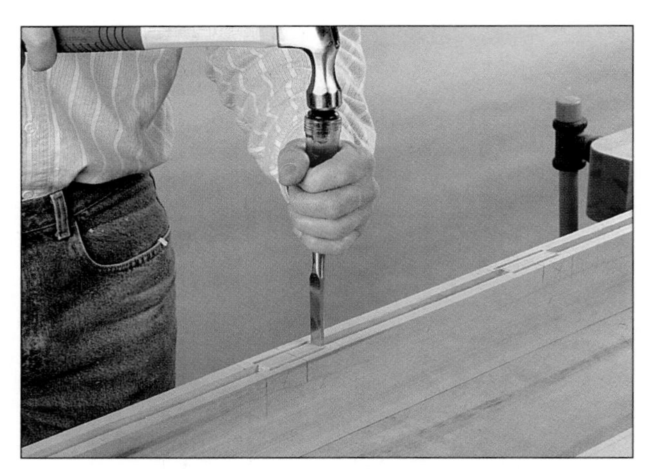

FIGURE E: After the grooves have been routed, disassemble the headboard and use a sharp chisel to square the rounded ends.

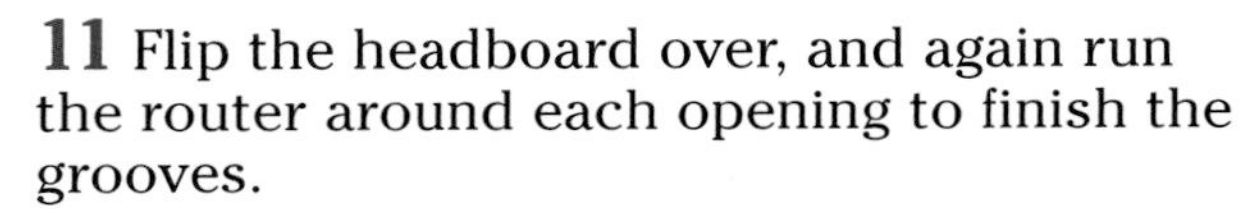

11 Flip the headboard over, and again run the router around each opening to finish the grooves.

12 Disassemble the headboard parts and use a sharp chisel to square the rounded slot ends left by the router *(see Figure E).* Cut the ½-in.-thick plywood panels, Part E, to size, lightly sand them and thoroughly dust them off.

13 Spread glue in the headboard joint slots and on all joining plates. Insert the plates and join the mullions to the wide rail. Slide the panels in place *(see Figure F)* and install the top rail. Position the stiles and clamp the assembly.

14 Compare opposite diagonal measurements to check that it's square. If the measurements are different, adjust the clamps until they're the same.

15 Cut a piece of ¾-in.-thick poplar to size for the headboard cap, Part F.

16 Secure the cap to the top of the headboard using glue and nails *(see Figure G).*

17 Set the nailheads below the surface and fill the holes with wood filler.

18 Sand the entire headboard with 120-, 150- and 180-grit sandpaper, dusting off the assembly thoroughly between grits. Carefully ease all sharp edges when you sand.

FIGURE F: Slide the panels into place, and join the top rail to the mullions. Finish the assembly by joining the stiles to the rail ends.

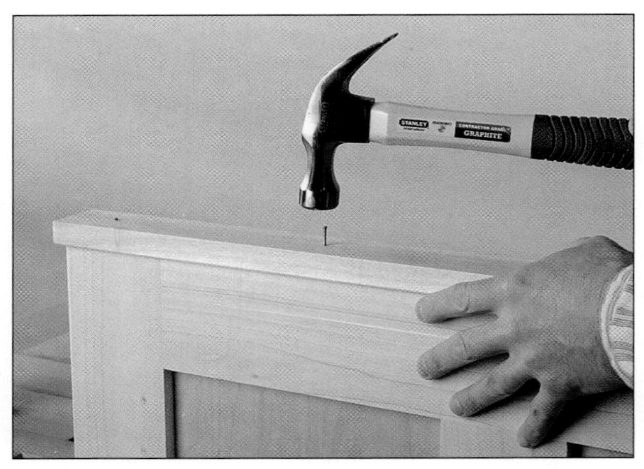

FIGURE G: Use glue and finish nails to attach the ¾-in.-thick poplar cap to the headboard. Set the nails and fill the nail holes.

Building the Storage Boxes

1 The main support for the bed is provided by two back-to-back storage box units. Use your circular saw guided by a straightedge to cut the plywood pieces, Parts G through J, for these boxes—a fine-tooth blade will minimize tearout *(see Figure H).*

2 Mark the locations of joining plate slots in the box parts.

3 Use your plate joiner to cut the slots. Clamp guides and fences to the panels and worktable to help register the plate joiner when making these cuts *(see Figure I).*

4 When you cut the slots in the end of a panel, you can use your worktable as the registration surface *(see Figure J).*

5 To minimize the need for clamps, some of the plate joints in the bed are used only to align the joint, and screws are used in place of glue. Use dry plates to assemble the partitions, back panels and top and bottom panels *(see Figure K).*

6 Then install screws to hold the parts together *(see Figure L).*

7 Spread glue in the plate slots and on the joining plates for the end pieces.

8 Position the ends and use 6d finish nails to hold the joints tight *(see Figure M).*

9 Rip strips of ¾-in. poplar to a width of ⅞ in. for the facing, Parts K through M.

10 Cut the pieces to length using a miterbox.

FIGURE H: To cut the plywood storage box pieces, guide your circular saw with a straightedge clamped to the workpiece.

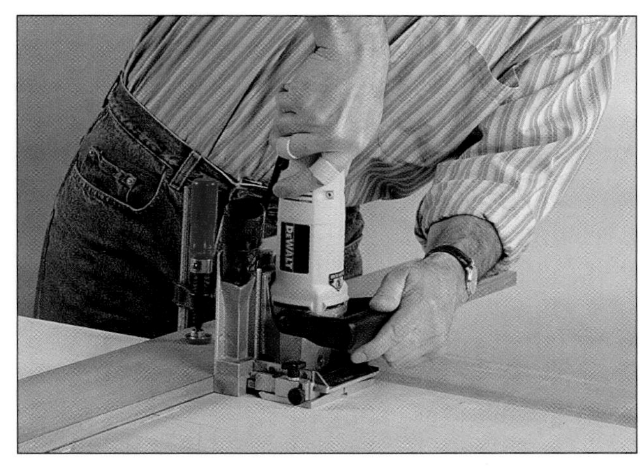

FIGURE I: Clamp a guide across the top and bottom panels of the storage boxes to help locate the plate joiner when cutting the slots.

FIGURE J: Use your worktable as a registration surface when cutting the slots in the storage box partition panels.

FIGURE K: Install the plates and position the partitions followed by the backs. Finally, place the top panel in position.

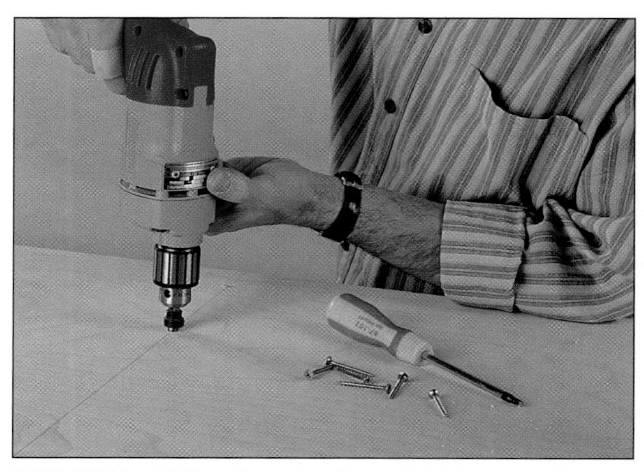

FIGURE L: With all parts aligned, bore screwholes through the top and secure the panel. Turn the assembly over and fasten the bottom.

FIGURE M: Use 6d finish nails in place of clamps to hold the ends tightly in place while the glue on the joining plates sets.

FIGURE N: Use glue and nails to fasten the facing strips to the storage boxes. Use a clamp to hold the strips while you drive the nails.

11 Apply the facing to the outside edges of the boxes *(see Figure N)* as shown in the Technical Illustration. Pay careful attention to the overhang of each facing strip.

12 Set and fill the nailheads.

Building the Toe Kick and Platform

1 Cut the pieces to size for the toe kick base, Parts N through P.

2 Apply glue to the joints and then use clamps to hold the pieces while you nail them together *(see Figure O).*

3 Use glue and finishing nails to secure the cleats.

4 Cut the mattress platform, Part Q, and poplar edges, Part R, and end, Part S, to size.

5 Cut the joining plate slots for fastening the 3½-in.-high edges and end to the plywood.

6 Apply glue to the joints and assemble the pieces *(see Figure P).* Use 6d finish nails to hold the joints tight.

Assembling and Finishing the Bed

1 Start assembling the bed by placing the two storage box units back to back.

2 Use clamps to hold the units together while you bore pilot holes and drive screws through the back of one unit into the back of the other.

3 Turn the box assembly upside down and position the toe kick base so there is a uniform setback on both sides and at the foot of the bed.

4 Fasten the toe kick by screwing through the cleats into the bottom of the box *(see Figure Q).*

5 Place the panel assembly right side up, position the headboard and temporarily clamp it in place.

6 Bore and countersink pilot holes and screw the headboard to the end of the base assembly *(see Figure R).*

7 Position the platform over the storage boxes, with its open end against the headboard.

8 Bore and countersink pilot holes, then fasten the platform with screws *(see Figure S).*

9 Inspect the bed and fill any remaining nail holes with wood filler.

10 Sand all bed parts, finishing with 180-grit sandpaper, and dusting off between grits.

11 After the final sanding of the barewood surfaces, completely dust the bed with a tack cloth.

12 Apply a good quality latex primer to all exposed surfaces.

13 When the primer is dry, lightly smooth the bed by hand, sanding with 180-grit paper to remove any surface imperfections.

14 Clean the bed again and follow with two coats of a good latex enamel, following the manufacturer's instructions.

FIGURE O: Assemble the toe kick base with glue and finish nails. Use clamps to keep the pieces from shifting as you nail.

FIGURE P: Join the platform and edges with plates and glue. Drive nails to hold the joints tight while the glue sets.

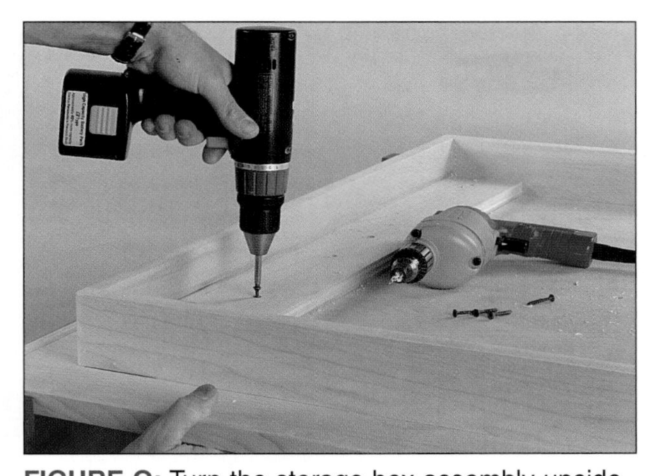

FIGURE Q: Turn the storage box assembly upside down and place the toe kick base over it. Secure the base with screws.

FIGURE R: Position the headboard and clamp it in place. Bore and countersink pilot holes and screw the headboard to the storage box.

FIGURE S: Place the platform assembly on the storage unit and tight against the headboard. Secure the platform with screws.

THE MUSCULAR SYSTEM
CHAMPION
Fun with Nature
JANSPORT

Kid's Room Dresser

While it may seem that some kids can go indefinitely wearing the same T-shirt and pair of jeans, most have an appetite for clothes that can strain the most ample budgets. And where do you put it all?

Well, before you decide to build an addition on the house, take a close look at our solution. Designed to match the rest of our bedroom suite *(see page 81)*, this dresser features four generous drawers that slide effortlessly on ball-bearing-equipped tracks, and sturdy plywood construction with solid poplar detailing.

Best of all, the dresser is easy to build. The case joinery utilizes a combination of joining plates and screws—the plates ensure perfect joint alignment while the screws provide holding power and eliminate the need for glue and a lot of long clamps. We've also streamlined drawer construction by employing simple and fast glue-and-nail joints.

As shown in the photo, we've accessorized our dresser with a wall-mounted storage unit. This piece is based on the shelf assembly featured with our desk *(see page 102)*. To build the wall-mounted unit, follow the instructions given for the desk unit, but eliminate the leg sections and cut the end panels to 10½ in. long.

Constructing the Dresser Case

1 Equip your circular saw with a fine-tooth blade to cut the plywood case sides, Part A, bottom, Part B, and cleats, Part C, to size. For accuracy, use a straightedge guide positioned at the appropriate distance from your cutline. With the guide square to the edge of the panel, hold the saw base against the guide while moving the saw slowly forward.

2 Use a household iron to apply veneer tape to the front edges of the sides, bottom and front cleat *(see Figure A)*. Position the 13⁄16-in.-wide tape so there's a slight overhang on each side of the panel. Set the iron to its highest setting, and advance it slowly while you press down firmly.

3 Trim the excess tape flush to the panel faces with a razor-sharp chisel. If the tape

FIGURE A: Heat the hot-melt adhesive on the back of the veneer tape with an iron. Slowly advance the iron with firm pressure.

FIGURE B: Use a sharp chisel to trim the overhanging edges of the veneer tape. If the tape begins to tear, reverse the cutting direction.

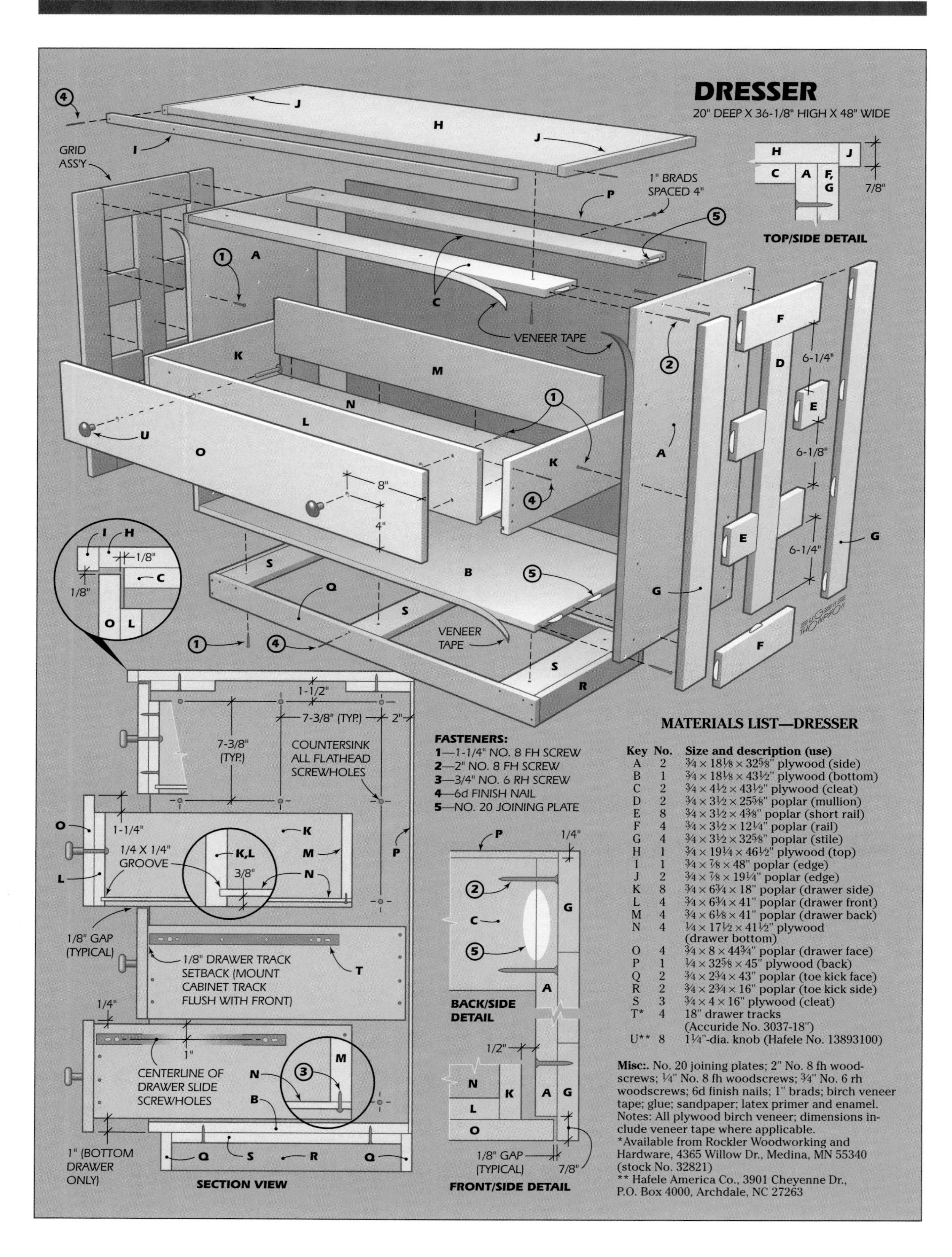

MATERIALS LIST—DRESSER

Key	No.	Size and description (use)
A	2	¾ × 18⅛ × 32⅝" plywood (side)
B	1	¾ × 18⅛ × 43½" plywood (bottom)
C	2	¾ × 4½ × 43½" plywood (cleat)
D	2	¾ × 3½ × 25⅝" poplar (mullion)
E	8	¾ × 3½ × 4⅜" poplar (short rail)
F	4	¾ × 3½ × 12¼" poplar (rail)
G	4	¾ × 3½ × 32⅝" poplar (stile)
H	1	¾ × 19¼ × 46½" plywood (top)
I	1	¾ × ⅞ × 48" poplar (edge)
J	2	¾ × ⅞ × 19¼" poplar (edge)
K	8	¾ × 6¾ × 18" poplar (drawer side)
L	4	¾ × 6¾ × 41" poplar (drawer front)
M	4	¾ × 6⅛ × 41" poplar (drawer back)
N	4	¼ × 17½ × 41½" plywood (drawer bottom)
O	4	¾ × 8 × 44¾" poplar (drawer face)
P	1	¼ × 32⅝ × 45" plywood (back)
Q	2	¾ × 2¾ × 43" poplar (toe kick face)
R	2	¾ × 2¾ × 16" poplar (toe kick side)
S	3	¾ × 4 × 16" plywood (cleat)
T*	4	18" drawer tracks (Accuride No. 3037-18")
U**	8	1¼"-dia. knob (Hafele No. 13893100)

Misc:. No. 20 joining plates; 2" No. 8 fh woodscrews; ¼" No. 8 fh woodscrews; ¾" No. 6 rh woodscrews; 6d finish nails; 1" brads; birch veneer tape; glue; sandpaper; latex primer and enamel. Notes: All plywood birch veneer; dimensions include veneer tape where applicable.

*Available from Rockler Woodworking and Hardware, 4365 Willow Dr., Medina, MN 55340 (stock No. 32821)

** Hafele America Co., 3901 Cheyenne Dr., P.O. Box 4000, Archdale, NC 27263

tends to tear, reverse the direction of the chisel *(see Figure B)*.

4 Mark the locations of joining plate slots on the plywood case parts.

5 First cut the slots in the ends of the cleats and bottom. For good joint registration, hold both the piece and plate joiner tight to your worktable.

6 To cut the slots in the case sides, first clamp a tall fence to the worktable. Use this fence as a support to hold the sides in a vertical position while cutting the slots *(see Figure C)*.

7 Bore screw clearance holes through the case sides and countersink the holes so that the screwheads will be slightly recessed.

8 Install joining plates in the case joints. Since these joints depend on screws for their strength, don't apply glue to the plates.

9 Assemble the case parts. Use clamps to hold the parts together while you bore pilot holes into the panel edges and drive the screws *(see Figure D)*. Set this assembly aside while you construct the poplar grids.

Making The Grids

1 Using a square to guide your circular saw, cut the poplar grid mullions, Part D, short rails, Part E, long rails, Part F, and stiles, Part G, to length.

2 Lay out the joining plate slot locations and cut the slots. When cutting into endgrain, especially on short pieces, clamp the part to the worktable *(see Figure E)*.

3 Apply glue to the slots and plates for the joints between the mullion and the short, center rails.

4 Join the rails to the mullion *(see Figure F)*, and then use clamps to pull the joints tight until the glue sets.

5 Join the top and bottom rails to the mullion ends. Again, clamp the joints *(see Figure G)*.

6 Finally, join the two stiles to the rail ends. Use a clamp at each rail to ensure that the joints are tight.

7 Compare opposite diagonal measurements of the grid assembly to be sure that it is square. If the measurements differ, adjust the clamps until they are the same.

FIGURE C: Use a plate joiner to make the slots for the case. Clamp case sides vertically and register the slots against your work surface.

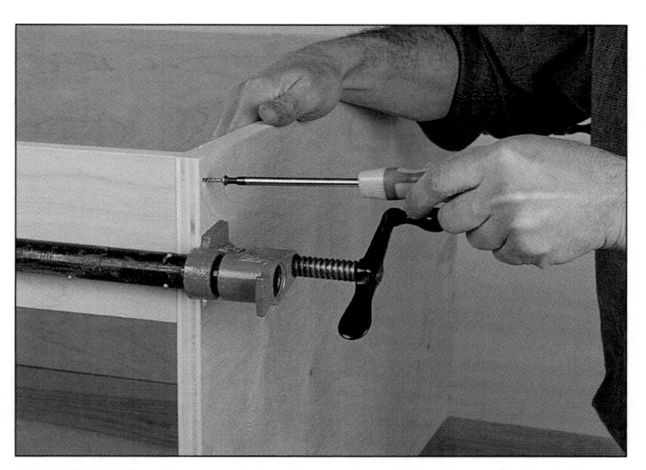

FIGURE D: Assemble the case by screwing the sides to the cleats and bottom. Use clamps to hold the pieces while you drive the screws.

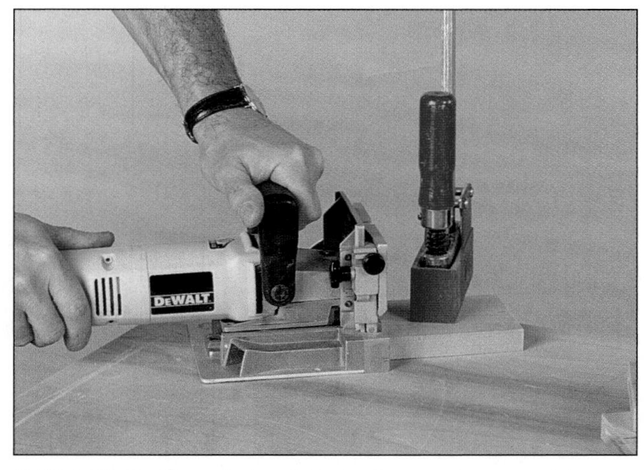

FIGURE E: Cut plate slots in the grid assembly components. Clamp pieces to your work surface for safe and accurate cuts.

FIGURE F: Spread glue in the slots and on the plates and assemble short rails to the mullion. Clamp the subassembly until the glue sets.

FIGURE G: Apply glue to the end rail joints and clamp to the mullion. After the glue has set, attach the grid stiles with glue and plates.

Making the Top and Assembling the Case

1 Cut the dresser top, Part H, to size.

2 Use your circular saw and rip guide to cut edge strips, Parts I and J, for the top, and then crosscut the strips to length.

3 Apply glue to one end of the top panel and position a strip so it's flush with the top surface and overhanging 1/8 in. on the bottom. Clamp the strip while you drive 6d finish nails. The clamp will keep the strip from moving, ensuring that it stays flush with the top of the panel.

4 Apply the strip at the opposite end of the panel, and add the front edge. Set the nailheads.

5 Position a grid assembly against each plywood case side, adjusting for the proper overhang at the front and back edges.

6 Use clamps to temporarily hold the parts together, and bore and countersink pilot holes through the case sides. Pay attention to the hole locations as shown in the technical illustration so you do not place screws where they might interfere with the drawer track installation.

7 Fasten the grids with 1 1/4-in. No. 8 fh screws *(see Figure H)*.

8 Clamp the top in place while you bore and countersink pilot holes through the cleats *(see Figure I)*.

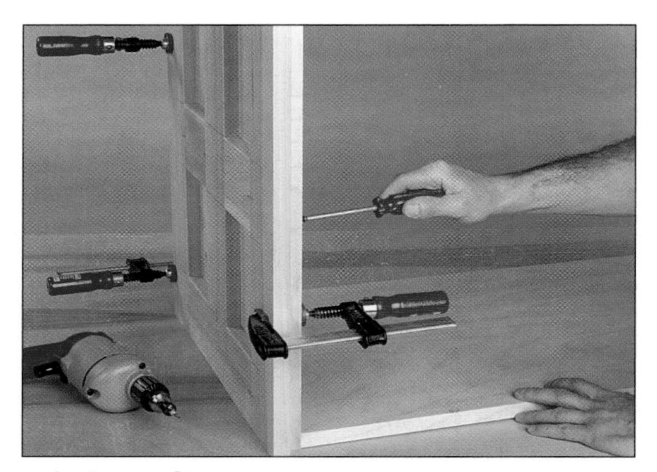

FIGURE H: Clamp the grid assemblies to the case sides. Bore and countersink screw pilot holes and drive screws to fasten the grids.

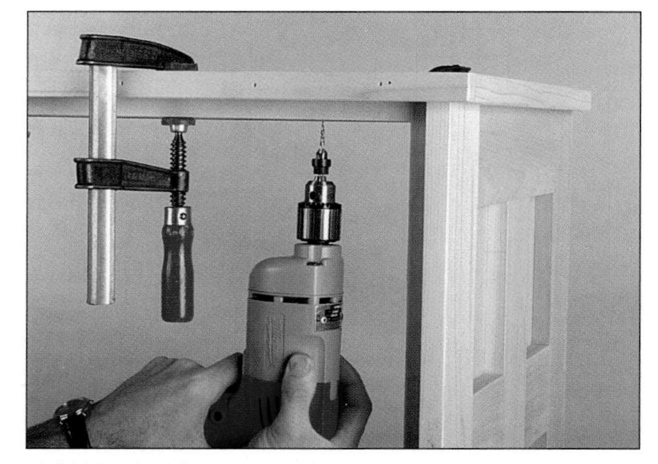

FIGURE I: After gluing the edges to the plywood top, clamp the top in place and bore screwholes. Then, screw the case to the top.

9 Fasten the top to the case with 1¼-in. No. 8 fh screws.

Building the Drawers

1 Rip and crosscut the drawer sides, Part K, fronts, Part L, and backs, Part M, to finished dimension.

2 Rout the grooves in the drawer sides and fronts for the bottom panels. Use a ¼-in. straight bit and an edge-guide accessory *(see Figure J).*

3 Sand the interior drawer surfaces with 120-, 150- and 180-grit sandpaper before assembly.

4 Dust off the pieces and assemble the drawers. Apply glue to the mating surfaces, and use clamps to hold the parts together while you drive 6d finish nails *(see Figure K).*

5 Cut the ¼-in. plywood drawer bottom panels, Part N, to size.

6 Sand the bottom panels, finishing with 180-grit sandpaper.

7 After dusting off the bottom panels, slide each one into the grooves of an assembled drawer.

8 Drill pilot holes and drive screws through the panels into the drawer backs to secure the panels *(see Figure L).* Glue or screws are not used in the panel grooves.

9 Mount the drawer rails (one half of the drawer track, Part T,) to the drawer sides

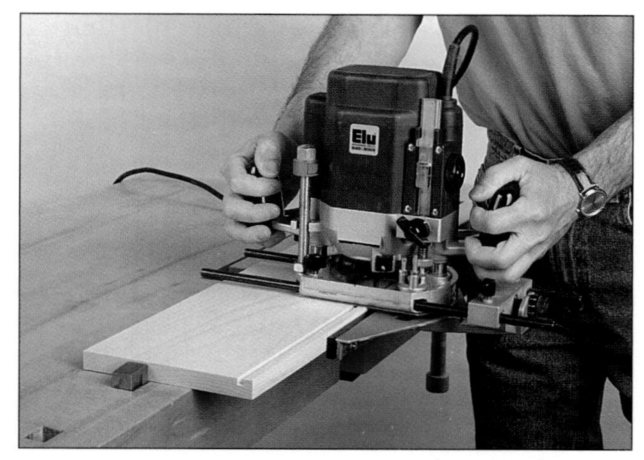

FIGURE J: Rout grooves for the drawer bottoms in the drawer front and side panels. Use a ¼-in. straight bit and router edge guide.

FIGURE K: Apply glue to the drawer box joints. Then clamp the boxes together and drive nails at the corners. Set the nailheads.

FIGURE L: After sliding the ¼-in. bottom panels in place, secure them to the drawer backs with ¾-in. No. 6 rh screws.

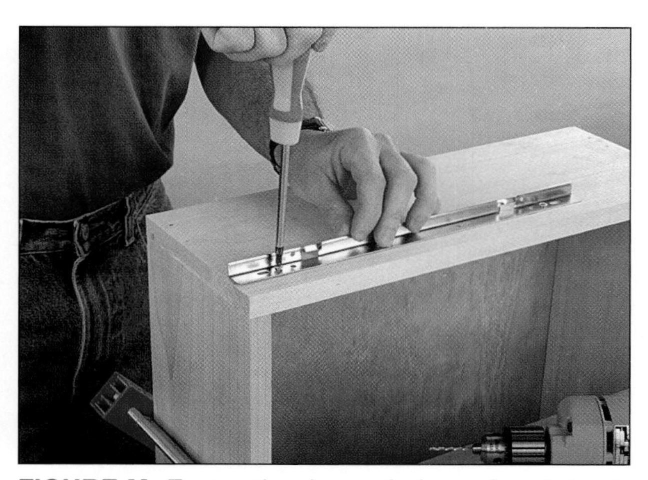

FIGURE M: Fasten the drawer halves of each track to the drawer sides with screws driven through the vertical slotted holes.

(see Figure M). Drive screws through the vertical slots at the ends of each rail to allow for adjustment after installation. The rails must be mounted ⅛ in. back from the front edge of the box.

10 Lay the case on its side on your worktable to mount the remaining track halves *(see Figure N)*. Drive screws through the horizontal slots at each end of the slides to allow for track adjustment. This time, position the track halves so that the front ends are flush with the front edge of the dresser case.

Assembling and Painting the Dresser

1 Cut the back panel, Part P, to size.

2 Compare opposite diagonal measurements of the case to be sure that it is square. Then nail the back in place with 1-in. brads *(see Figure O)*.

3 To install the drawers, engage the drawer rail under the small hooks at the back edge of the tracks, then lower the drawer over the plastic clips until you hear them click into place *(see Figure P)*.

4 Cut poplar stock to size for the drawer faces, Part O.

5 Starting with the bottom drawer, clamp the face to the box, bore screwholes from inside and drive screws to fasten the face *(see Figure Q)*.

6 After all faces are installed, check their alignment and adjust the slides to achieve a uniform ⅛-in. space around each.

7 Drive screws into all the remaining track-mounting holes.

8 Bore holes for the drawer knobs, Part U, and mount the knobs with 1¾-in.-long mounting screws.

9 Remove the drawers for painting.

10 Cut the toe kick base faces, Part Q, sides, Part R, and cleats, Part S, to size.

11 Use glue and 6d finish nails to assemble the toe kick base.

12 Lay the cabinet on its back, and clamp the base to the case bottom while you bore pilot holes and screw the toe kick in place

FIGURE N: Install the case tracks by screwing through the horizontal slotted holes. Place tracks flush with the front edges of case.

FIGURE O: After checking that the case is square, install the ¼-in. back panel with 1-in. brads driven about 4 in. apart.

FIGURE P: Install each drawer by engaging the drawer member with the hook at the back of the track. Then, lower front of drawer.

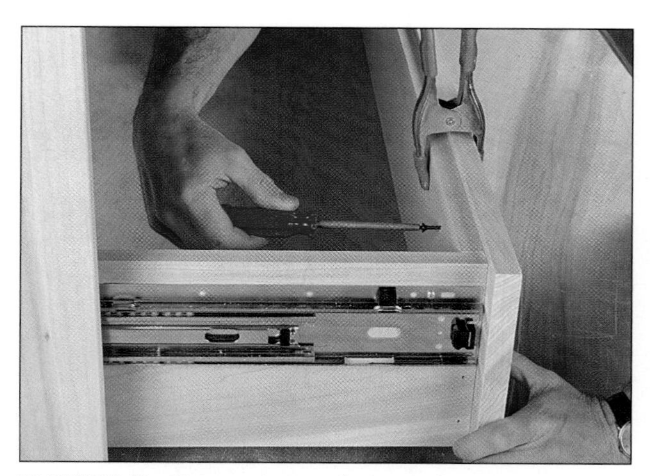

FIGURE Q: Clamp drawer face to drawer box. Then bore and countersink screw pilot holes. Attach each face and check alignment.

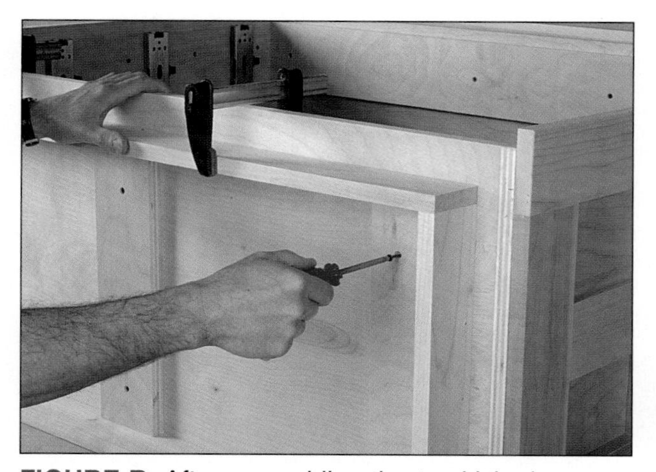

FIGURE R: After assembling the toe kick pieces with glue and nails, clamp the base to the dresser, bore screwholes and drive screws.

(see Figure R).

13 Set all nailheads and fill the holes with wood filler.

14 Sand the case and drawer parts, finishing with 180-grit paper, and clean away the dust.

15 Apply a good latex primer to all cabinet surfaces and drawer faces. If you wish to finish the drawer boxes, use two coats of shellac.

16 When the primer is dry, lightly sand all surfaces with 180-grit paper. Then dust it off before painting.

17 Apply two coats of a quality latex semi-gloss enamel, following the directions supplied by the manufacturer.

18 When the paint is dry, reinstall the drawers.

Text and Photos by Neal Barrett
Technical Illustration by Eugene Thompson

CLASSIC STRESS RESPONSE
I CAN'T TAKE IT!!
Baden
Student
Macs For Dummies
INTERNET
iMac
CARTOPEDIA
SPACE & UNIVERSE
Thinkin' Things
WORLD BOOK
DORLING KINDERSLEY
WORLD REFERENCE ATLAS
PRELUTSKY
a PIZZA the size of the SUN
Microsoft
ENCARTA 98 ENCYCLOPEDIA

Kid's Desk

Kids are not known for their neat work habits—neither are many adults, for that matter. But at least with kids, there's always hope that they can learn something better. Certainly the right desk can help.

Our design, matching the rest of the kid's room furniture (see page 81), provides sufficient work surface and storage space. The desktop is large enough to accommodate a computer, and it still has room for software, papers and books. The pedestal provides a drawer for smaller loose items, and two deep shelves can hold a small printer and other supplies. The storage shelf has lots of space for CDs, tapes and collectibles. If you want drawers on the desk instead of the open compartments we show, then simply use the instructions for the dresser drawers (page 93) to help you along your way.

The desk is built from birch plywood with poplar edge-banding, and the parts are held together with joining plates, nails, screws and glue. Any beginner can build it. Its construction is so rugged, it's just about impossible to damage it, and it disassembles to make it easier to move—a handy feature when transporting it from your shop to the bedroom.

Making Pedestal Parts

1 Rip and crosscut the plywood sides, Part A, shelves, Part B, and cleats, Part C, to size. Guide the circular saw using a straightedge clamped to the panel *(see Figure A)*. Note that a 40-tooth thin-kerf, crosscut blade was used for these cuts.

2 Use a clothes iron to apply birch veneer edge tape to the plywood pieces *(see Figure B)*.

3 Let the veneer cool to room temperature before trimming it to length and width using a sharp chisel *(see Figure C)*. If the veneer tears out as it is trimmed, cut from the opposite direction.

4 Mark the locations of the joining plate slots in the pedestal parts. The joining plates hold the parts in position, and screws are used to pull the parts together. The screwheads are hidden under the applied grids.

5 Cut the joining plate slots in the pedestal parts using the plate joiner. For the slots at the top and bottom of the pedestal sides, clamp a tall fence to the workbench. Then clamp a pedestal side to the fence, and cut the slots with the plate joiner held against the workbench *(see Figure D)*.

6 To cut the slots in the center of a panel, clamp a straightedge across the panel to guide the plate joiner *(see Figure E)*.

7 To cut the slots in the shelf ends and

FIGURE A: To cut the plywood desk parts to size, use a circular saw guided by a straightedge clamped to the panel's face.

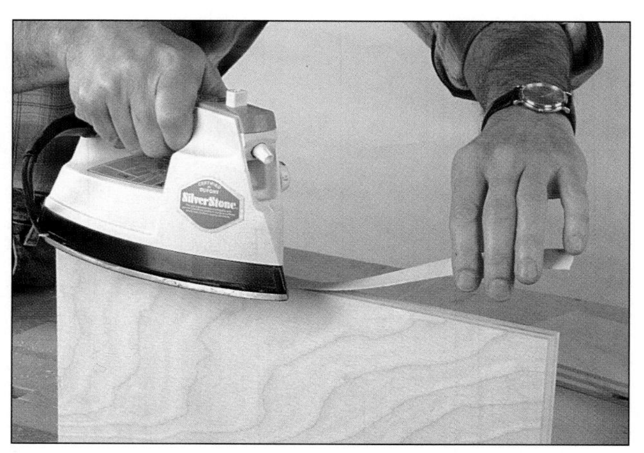

FIGURE B: Use an iron to apply heat to the birch veneer tape. The banding has heat-sensitive adhesive on its back.

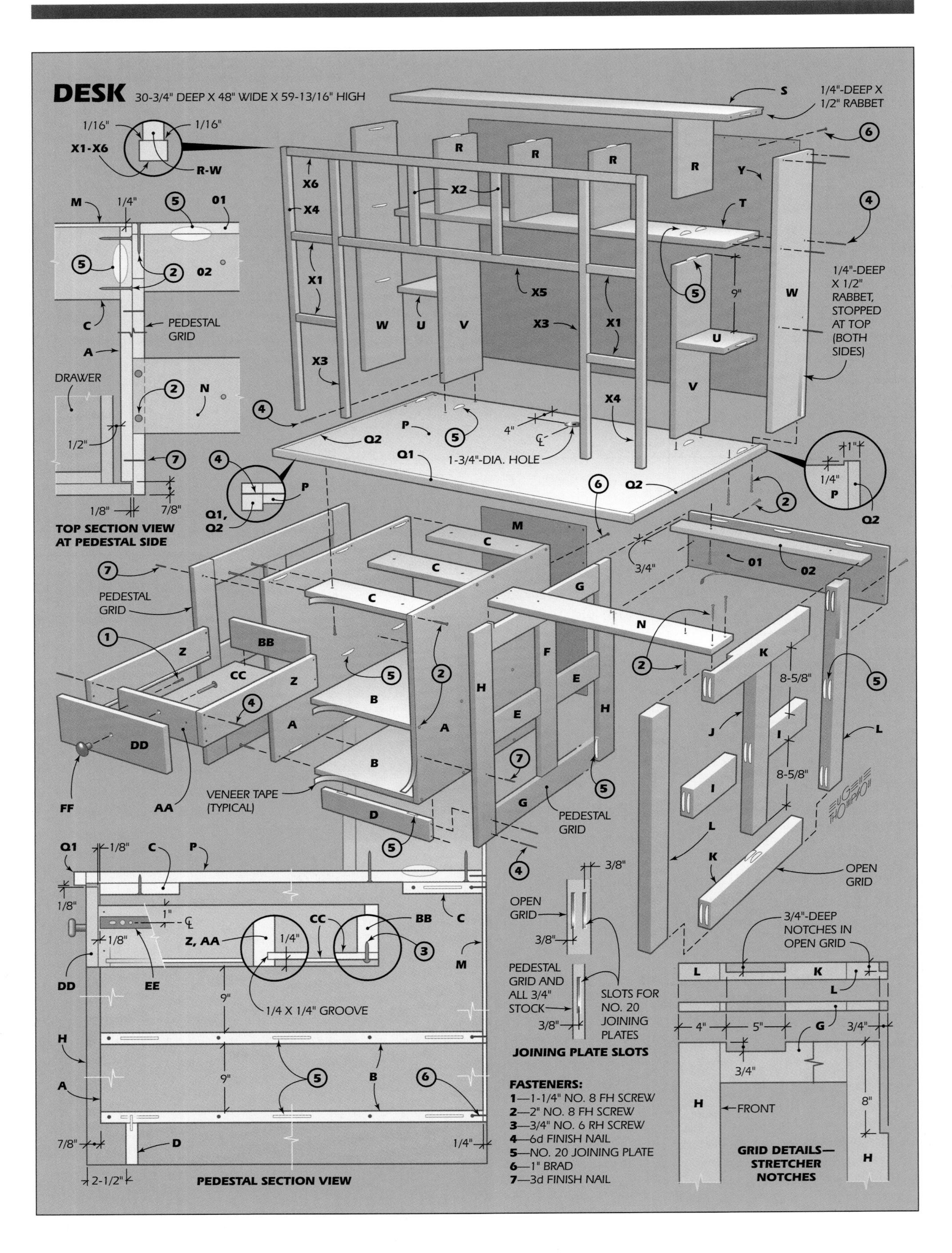
DESK 30-3/4" DEEP X 48" WIDE X 59-13/16" HIGH
1/4"-DEEP X 1/2" RABBET
1/4"-DEEP X 1/2" RABBET, STOPPED AT TOP (BOTH SIDES)
1-3/4"-DIA. HOLE
TOP SECTION VIEW AT PEDESTAL SIDE
PEDESTAL GRID
DRAWER
VENEER TAPE (TYPICAL)
OPEN GRID
3/4"-DEEP NOTCHES IN OPEN GRID
1/4 X 1/4" GROOVE
PEDESTAL SECTION VIEW
JOINING PLATE SLOTS
PEDESTAL GRID AND ALL 3/4" STOCK
SLOTS FOR NO. 20 JOINING PLATES
FASTENERS:
1—1-1/4" NO. 8 FH SCREW
2—2" NO. 8 FH SCREW
3—3/4" NO. 6 RH SCREW
4—6d FINISH NAIL
5—NO. 20 JOINING PLATE
6—1" BRAD
7—3d FINISH NAIL
GRID DETAILS—STRETCHER NOTCHES
FRONT

cleats, hold the workpiece to the bench, and use the top as the registration surface *(see Figure F)*.

8 Bore and countersink pilot holes through the pedestal sides for joining the sides, shelves and cleats.

9 Install joining plates in the side panels, and check the pieces' fit before the assembly sequence *(see Figure G)*.

MATERIALS LIST—DESK

Key	No.	Size and description (use)
A	2	¾ × 25 × 28⅞" plywood (side)
B	2	¾ × 15 × 28⅞" plywood (shelf)
C	3	¾ × 5 × 15" plywood (cleat)
D	1	¾ × 2¾ × 16½" poplar (toe kick)
E	4	¾ × 3½ × 9¾" poplar (short rail)
F	2	¾ × 3½ × 20¾" poplar (mullion)
G	4	¾ × 3½ × 23" poplar (rail)
H	4	¾ × 3½ × 27¾" poplar (stile)
I	2	1½ × 3½ × 9¾" poplar (short rail)
J	1	1½ × 3½ × 20¾" poplar (mullion)
K	2	1½ × 3½ × 23" poplar (rail)
L	2	1½ × 3½ × 27¾" poplar (stile)
M	1	¼ × 16½ × 25" plywood (back)
N	1	¾ × 5 × 28½"plywood (cleat)
O1	1	¾ × 8 × 28½" plywood (back cleat)
O2	1	¾ × 3 × 27" plywood (cleat)
P	1	¾ × 30 × 46½" plywood (desktop)
Q1	1	¾ × ⅞ × 48" poplar (edging)
Q2	2	¾ × ⅞ × 30" poplar (edging)
R	4	¾ × 8 × 9" plywood (partition)
S	1	¾ × 8¼ × 45" plywood (top)
T	1	¾ × 8 × 45" plywood (shelf)
U	2	¾ × 6 × 8" plywood (shelf)
V	2	¾ × 8 × 20¾" plywood (side)
W	2	¾ × 8¼ × 31¼" plywood (side)
X1	4	¾ × ⅞ × 5⅞" poplar (facing)
X2	2	¾ × ⅞ × 8⅞" poplar (facing)
X3	2	¾ × ⅞ × 30 7/16" poplar (facing)
X4	2	¾ × ⅞ × 31 5/16" poplar (facing)
X5	1	¾ × ⅞ × 31⅜" poplar (facing)
X6	1	¾ × ⅞ × 44⅞" poplar (facing)
Y	1	¼ × 31¾ × 46" plywood (back)
Z	2	¾ × 4 × 18" poplar (drawer side)
AA	1	¾ × 4 × 12½" poplar (drawer front)
BB	1	¾ × 3½ × 12½" poplar (drawer back)
CC	1	¼ × 13 × 17½" plywood (drawer)
DD	1	¾ × 5¼ × 14¾" poplar (drawer face)
EE*	1	18" drawer tracks (Accuride No. (3037-18")
FF**	1	1¼"-dia. knob (Hafele No. 13893100)

Misc. No. 20 joining plates; 2" No. 8 fh woodscrews; 1¼" No. 8 fh woodscrews; ¾" No. 6 rh woodscrews; 6d and 3d finish nails; 1" wire brads; birch veneer (Rockler No. 10801); grommet (Rockler No. 14598); glue; sandpaper; latex primer and enamel.

Notes: All plywood birch veneer dimensions include veneer tape where applicable.
*Available from Rockler Woodwork and Hardware, 4365 Willow Dr., Medina, MN 55340 (stock No. 32821)
**Hafele America Co., 3901 Cheyenne Dr., P.O. Box 4000, Archdale, NC 27263

FIGURE C: Use a chisel to trim the birch veneer tape to width. Cut from the opposite direction if the tape tears out.

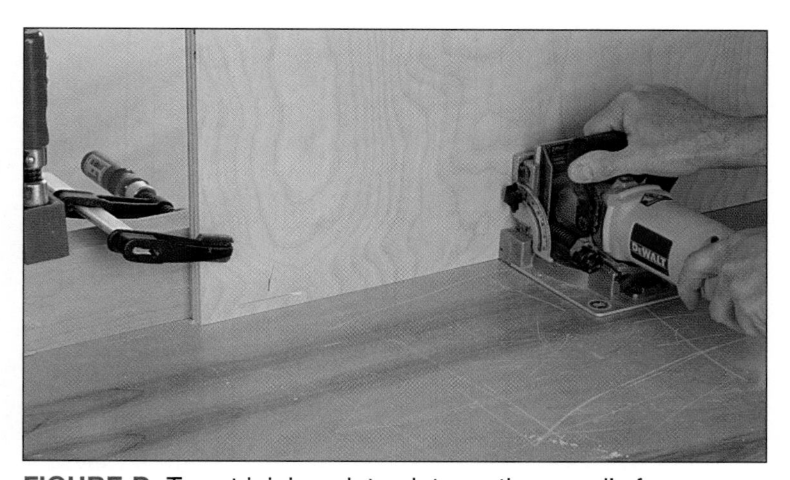

FIGURE D: To cut joining plate slots on the panel's face, clamp the panel upright, and slide the plate joiner on the work surface.

FIGURE E: To cut plate slots in the panel's center, clamp a fence across the panel to guide the vertically positioned plate joiner.

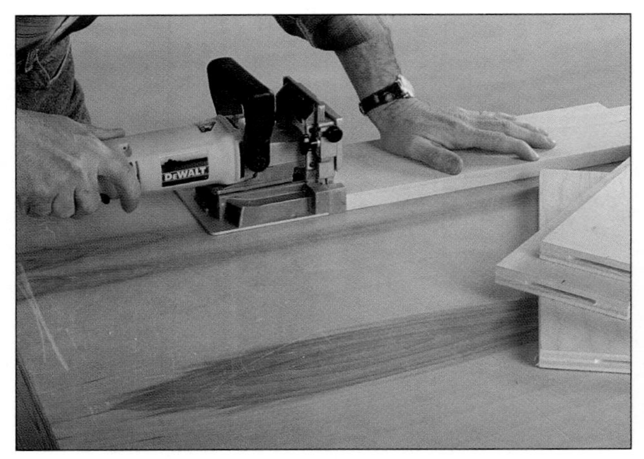

FIGURE F: To cut plate slots in the end of plywood parts, hold the workpiece down and slide the plate joiner on the bench.

FIGURE G: Bore and countersink the pilot holes in the pedestal sides, and test fit the parts before the final assembly.

Constructing the Desk

1 Assemble the pedestal sides, shelves and cleats without using glue in the plate joints.

2 Bore pilot holes into the ends of the shelves and cleats. Then drive the screws to fasten the sides to these parts *(see Figure H)*.

3 Rip and crosscut the poplar toe kick, Part D, to size.

4 Cut the joining plate slots in its top edge.

5 Spread glue in the slots and on the plates. Clamp it in place until the glue sets.

6 Rip and crosscut all the grid pieces: Parts E through H for the pedestal grid and Parts I through L for the open grid.

7 Mark the parts for joining plates. Use two plates at each joint on the open grid.

8 Cut the plate slots. When cutting slots in the endgrain of the poplar pieces, clamp the workpiece to the bench.

9 To assemble either grid, first glue and clamp together the crosspieces in the center of the grid *(see Figure I)*. Spread glue on the joining plates and in the plate slots.

10 When the glue is dry, glue and clamp the horizontal pieces to the top and bottom of the cross, and then glue and clamp the two vertical pieces to the assembly.

11 Cut the cleat notch on the top of the pedestal grid and the open grid.

12 Cut the ¾-in.-deep rabbet on the open

FIGURE H: Clamp the pedestal parts together, bore pilot holes into the shelves and cleats, and drive the screws.

FIGURE I: Begin the assembly of the open grid by gluing and clamping together the center horizontal and vertical pieces.

FIGURE J: Cut the stopped rabbet in the open grid using a router. Then cut the end of the rabbet square using a chisel.

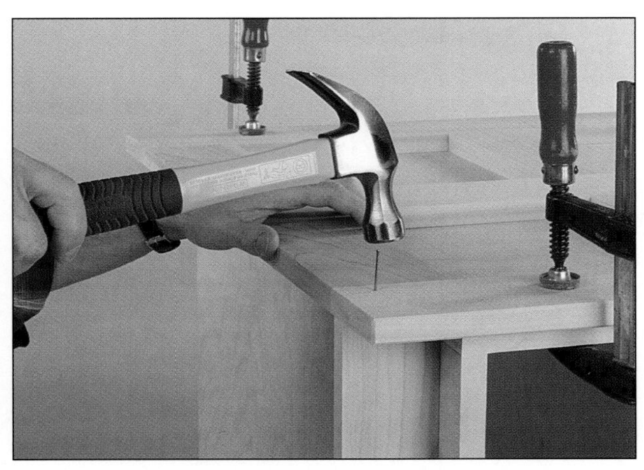

FIGURE K: Clamp the grid to the desk pedestal. Nail the grid to the pedestal sides with 3d nails and the toe kick with 6d nails.

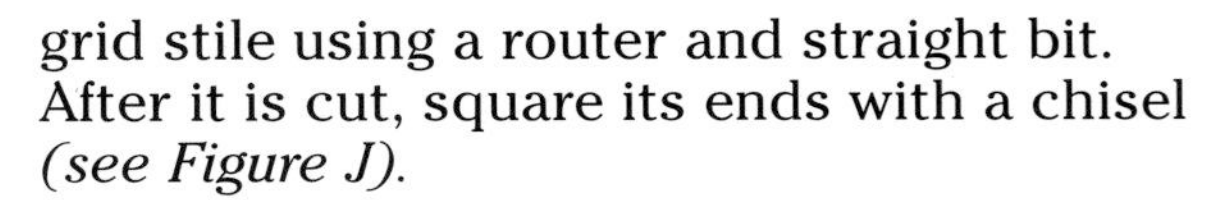

grid stile using a router and straight bit. After it is cut, square its ends with a chisel *(see Figure J)*.

13 Place the pedestal on its side, and position one of the pedestal grids on it. Clamp the grid to the pedestal, and nail it in place without using glue.

14 Nail the grid to the toe kick *(see Figure K)*.

15 Cut out the pedestal back, Part M, and nail it in place.

16 Cut the three cleats, Parts N, O1 and O2, that join the pedestal and open grid to size.

17 Apply birch veneer edge tape to the exposed edges of the front and back cleats.

18 Cut joining plate slots in the vertical back cleat and in the edge of the rear cleat.

19 Install joining plates with glue and clamp these cleats together. This cleat assembly is installed when the pedestal and open grid are joined.

20 Cut the desktop, Part P, to size.

21 Rip and crosscut the poplar edges, Parts Q1 and Q2, for the desktop.

22 Glue and clamp them to the top, and fasten them with 6d finish nails so they are flush with the top, but overhang the bottom by 1/8 in.

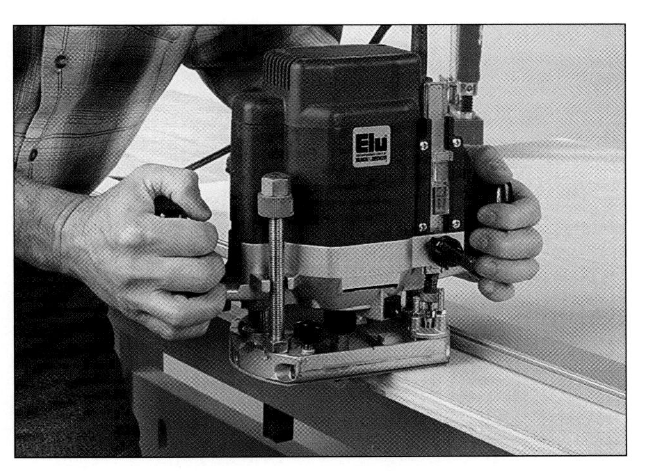

FIGURE L: Use a router to cut the notch in the top of the desk. The first cut forms a rabbet, and the second cut completes the notch.

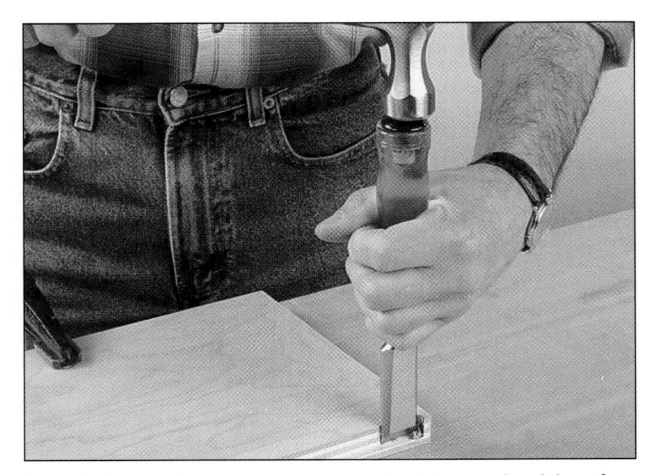

FIGURE M: Cut a stopped rabbet in each side of the storage shelf. Then use a chisel to cut the end of the rabbet square.

Building The Storage Shelf

1 Use a router and straight bit to cut the notch in the back of the desktop *(see Figure L)*. Square the ends of the cut with a chisel.

2 Bore the grommet hole in the top.

3 Cut the storage shelf pieces, Parts R through W, to size.

4 Apply birch veneer tape to the top edges of the storage shelf sides.

5 Use the router with an edge guide to cut the stopped rabbet on the sides. Then square the rabbet using a chisel *(see Figure M)*.

6 Cut the joining plate slots for the storage shelf.

7 Begin the storage shelf assembly by spreading glue in the slots and on the plates for the joints between the partitions and the top. Join these parts, and drive 6d finish nails to fasten the joints *(see Figure N)*. Clamp these joints until the glue sets.

8 Glue and clamp the short shelves to the inner sides until the glue sets.

9 Glue and clamp those sides to the bottom of the middle shelf *(see Figure O)*.

10 Glue and clamp the top and shelf assembly, then drive finish nails through the shelf into the inner side *(see Figure P)*.

11 Cut the storage shelf back, Part Y, to size and nail it in place.

12 Cut the poplar facing pieces, Parts X1 through X6, to size.

13 Glue and nail the poplar facing in place.

14 Cut the joining plate slots in the desktop to locate the storage shelf.

15 Install the plates with glue, then place the storage shelf on the desktop. Screw the desktop to the storage shelf *(see Figure Q)*.

FIGURE N: Begin the shelf assembly by joining the top and partitions with joining plates and glue. Also nail the pieces together.

FIGURE O: Join the short shelves and inner sides. Then use glue, plates and clamps to join these parts to the middle shelf.

FIGURE P: While holding the parts in position with clamps, drive a nail through the shelf and into the top of the inner side.

FIGURE Q: Fasten the storage shelf to the desktop by driving screws through the desktop and into the shelf sides.

FIGURE R: Spread glue on the endgrain joints of the drawer box parts, and clamp them together. Also fasten them with finish nails.

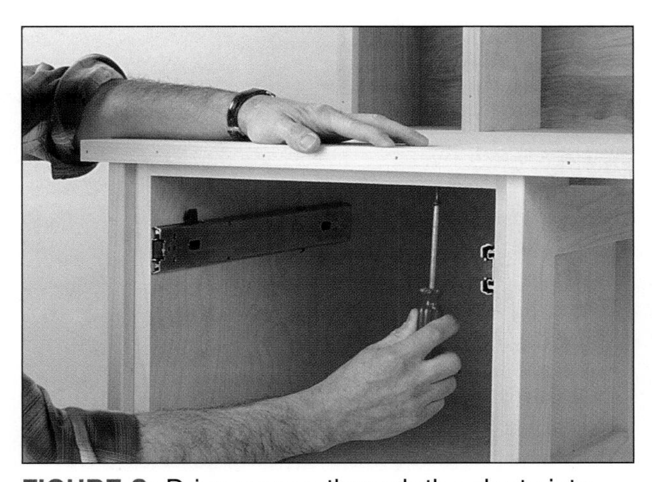

FIGURE S: Drive screws through the cleats into the bottom of the desk top to fasten the top and bottom subassemblies.

Assembling and Finishing the Desk

1 Cut the drawer sides, Part Z, front, Part AA, and back, Part BB, to size.

2 Clamp the parts together, and use glue and 6d finish nails to fasten them *(see Figure R).*

3 Cut the drawer bottom, Part CC, to size and install it.

4 Install the drawer tracks, Part EE, on the pedestal sides and the drawer sides. Then test fit the drawer in the pedestal.

5 Cut the drawer face, Part DD, to size, and screw it to the drawer box.

6 Install the drawer pull, Part FF.

7 Use screws, but not glue, to fasten the cleats between the open grid and the pedestal.

8 Place the desktop shelf assembly over the base. Then bore and countersink pilot holes through the cleats and into the bottom of the desktop.

9 Screw the desktop and storage shelf to the base *(see Figure S).*

10 To finish the desk, set all finish nails, and fill the holes.

11 Sand all the surfaces using 180-grit sand-paper and ease the edges.

12 Apply a coat of latex primer.

13 Lightly sand the primer.

14 Apply two coats of latex semigloss paint.

Text and Photos by Neal Barrett

Technical Illustration by Eugene Thompson

Kid's Storage Shelf

Text and Photos by Neal Barrett
Technical Illustration by Eugene Thompson

This storage shelf is designed to be as versatile as possible. In other words, it holds just about anything that is likely to end up scattered all over the floor in your child's room. Its tall spaces hold oversize children's books, school notebooks, stacks of games or stuffed animals. The smaller spaces are proportioned for tapes, CDs, art supplies and the odds and ends that inevitably clutter a child's room. Designed to match the other kid's room furniture *(see page 81)*, it also functions as a night stand because it falls at the right height for a lamp, radio and alarm clock. The shelf size can be changed easily to fit the layout of any room.

The construction of this piece employs the same materials as the child's bed—birch plywood and solid poplar. The assembly techniques rely on a combination of plate joints, screws and finish nails to draw shelf parts tightly together, so you won't need a bunch of expensive clamps.

Making Case Parts

1 Use a circular saw and 40-tooth thin-kerf, crosscut blade to cut the plywood case pieces, Parts A through E, to size. When plywood is cut, there is a tendency for the face veneer to chip where the blade exits the cut. You can prevent this chipping by using two techniques. First, clamp a straight board across the panel stock to guide the saw. Next, advance the saw slowly, and keep the saw base tight to the guide strip *(see Figure A)*.

2 Set up the router with a straight bit and an accessory edge guide. Adjust the router to cut the rabbet at the back edge of the case sides. Test the setup on a piece of scrap stock.

3 Clamp a case side to the workbench and cut the rabbet *(see Figure B)*. If you use a router bit with a ½-in.-dia. shank, you can make the cut in one pass. If you are using a bit with a ¼-in.-dia. shank, you should take two passes to cut the rabbet.

4 Mark the locations of plate joint slots in the cabinet sides, shelves and partitions. Note that the middle shelf has staggered slots on the top and bottom surfaces. It's important to stagger the slots to prevent too much wood from being removed in one location.

5 Clamp a guide block to the case sides and shelves to help locate the plate joiner when cutting the slots in the center of a panel *(see Figure C).*

6 When you cut the slots in the sides for the case top and bottom, you can use the fence on the plate joiner to register the cuts *(see Figure D).*

7 Use the workbench top as the registration surface when you cut the slots in the ends of the shelves and partitions. Firmly hold both the plate joiner and the workpiece to the benchtop when making the cut. Keep your fingers well away from the cutting area to avoid accidents.

8 Countersink pilot holes through the top, bottom and middle shelves *(see Technical Illustration for hole locations).*

Assembling the Case

1 Begin the case assembly process by joining the case top to the short partitions *(see Figure E).* Install the joining plates in their slots and position the short partitions over them. You do not need to use glue on these plates because they merely locate the joint.

2 Turn the assembly over, and bore pilot holes in the partitions *(see Figure F).* Then screw the partitions to the top panel.

FIGURE A: Clamp a straightedge across the workpiece and crosscut it with a circular saw. Support the piece that will be cut off.

FIGURE B: Use a straight bit in the router and the edge guide attachment to cut a rabbet along the back edge of the side panels.

FIGURE C: Clamp a fence across a case side, and use it to guide the plate joiner when cutting the plate slots for the shelves.

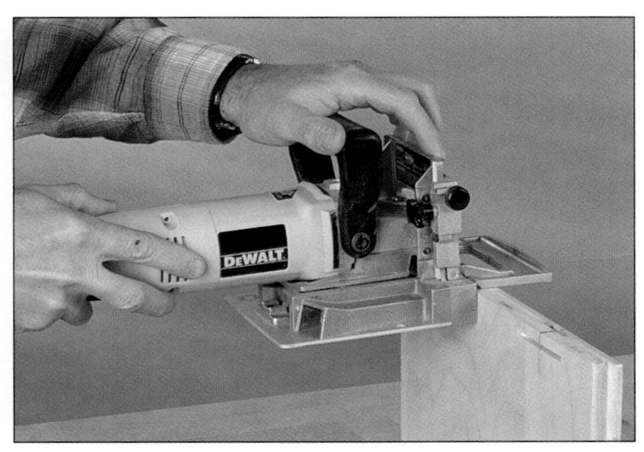

FIGURE D: Clamp the case sides upright in a vise and cut the slots along their upper edge using the plate joiner's fence for alignment.

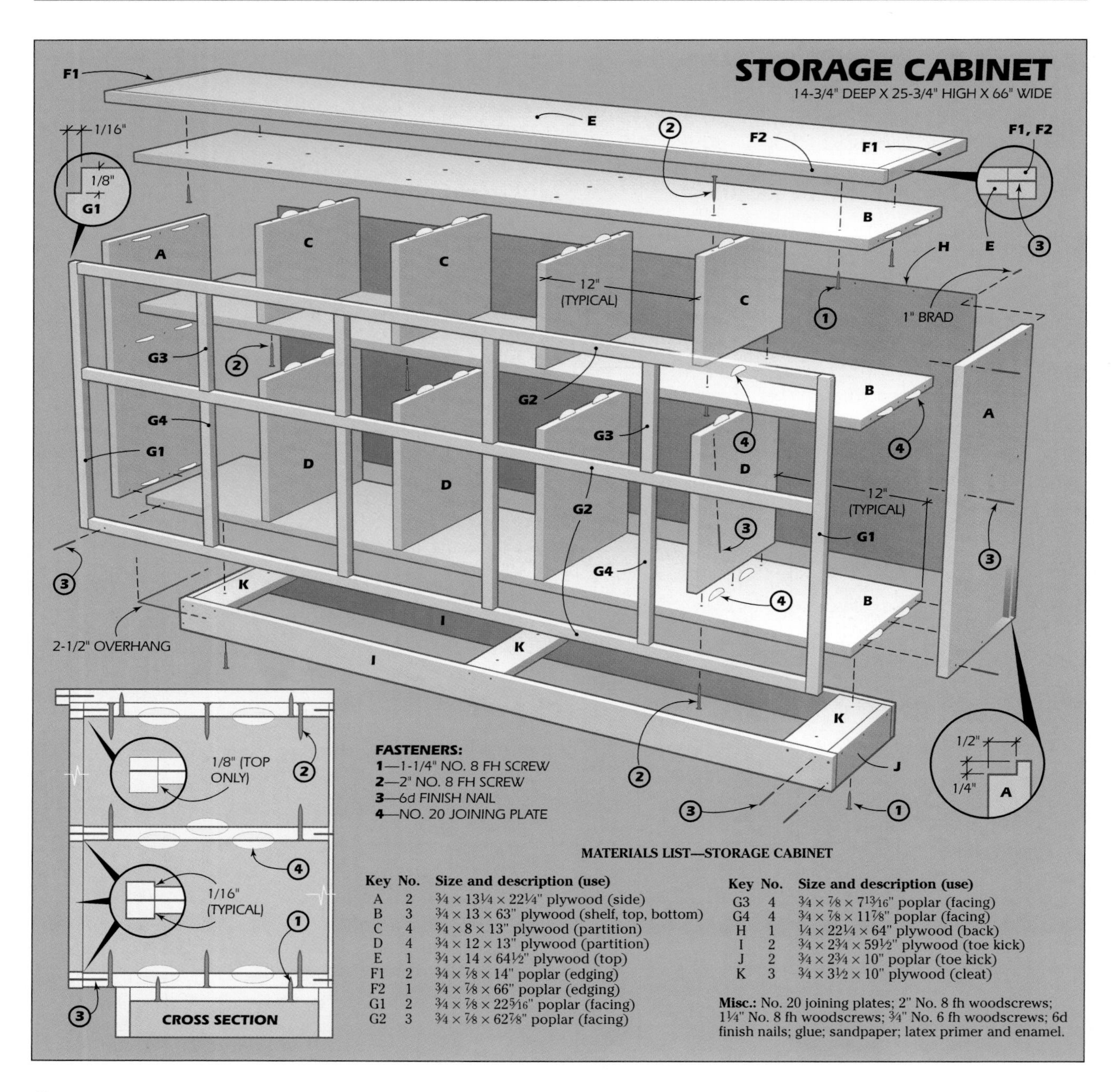

MATERIALS LIST—STORAGE CABINET

Key	No.	Size and description (use)
A	2	¾ × 13¼ × 22¼" plywood (side)
B	3	¾ × 13 × 63" plywood (shelf, top, bottom)
C	4	¾ × 8 × 13" plywood (partition)
D	4	¾ × 12 × 13" plywood (partition)
E	1	¾ × 14 × 64½" plywood (top)
F1	2	¾ × ⅞ × 14" poplar (edging)
F2	1	¾ × ⅞ × 66" poplar (edging)
G1	2	¾ × ⅞ × 22 5/16" poplar (facing)
G2	3	¾ × ⅞ × 62⅞" poplar (facing)
G3	4	¾ × ⅞ × 7 13/16" poplar (facing)
G4	4	¾ × ⅞ × 11⅞" poplar (facing)
H	1	¼ × 22¼ × 64" plywood (back)
I	2	¾ × 2¾ × 59½" plywood (toe kick)
J	2	¾ × 2¾ × 10" poplar (toe kick)
K	3	¾ × 3½ × 10" plywood (cleat)

Misc.: No. 20 joining plates; 2" No. 8 fh woodscrews; 1¼" No. 8 fh woodscrews; ¾" No. 6 fh woodscrews; 6d finish nails; glue; sandpaper; latex primer and enamel.

3 Spread glue in the joining plate slots for the joints between the short partitions and the middle shelf.

4 Place the middle shelf over the short partitions, bore pilot holes into the partition ends and fasten the shelf and partitions with screws.

5 Install joining plates in the slots for the joints between the bottom and the tall partitions.

6 Assemble the partitions and bottom, and fasten them with screws.

7 Spread glue in the slots and on the plates for the joints between the tall partitions and the middle shelf. Install the plates and clamp the assembly together.

8 Drive 6d finish nails through the middle shelf into the short partitions *(see Figure G).*

9 Spread glue in the slots and on the joining plates for the joints between the middle shelf, top and bottom, and the case sides. Assemble the parts, and drive 6d finish nails to fasten the joints.

FIGURE E: The short partitions are attached to the panel above with screws, so there is no need to use glue with the joining plates.

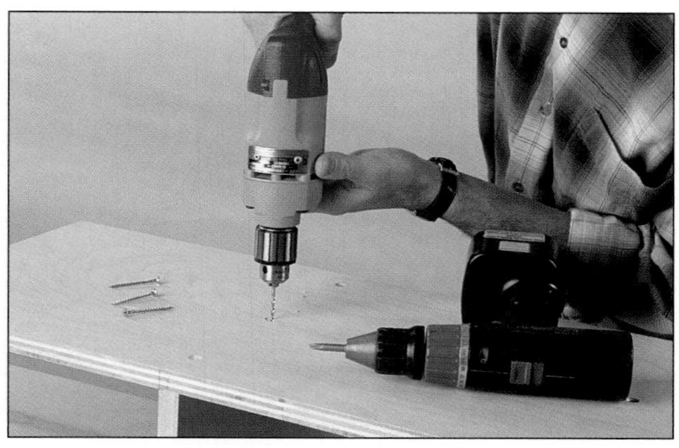

FIGURE F: Bore and countersink pilot holes into the top of the short partitions. Then drive screws to fasten the partitions and panel.

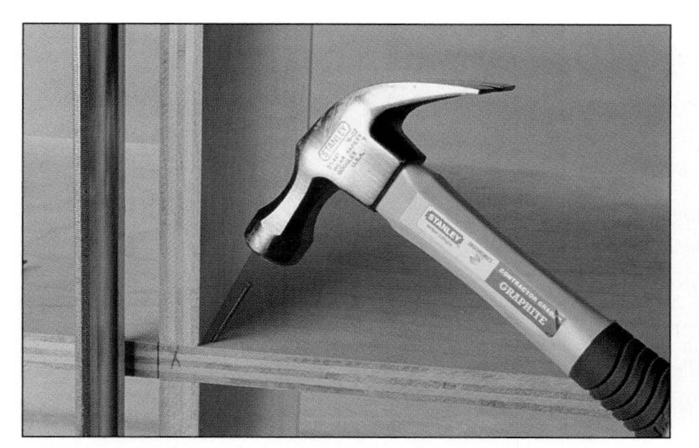

FIGURE G: Drive finish nails at an angle through the tall partitions and the middle shelf, and into the short partitions.

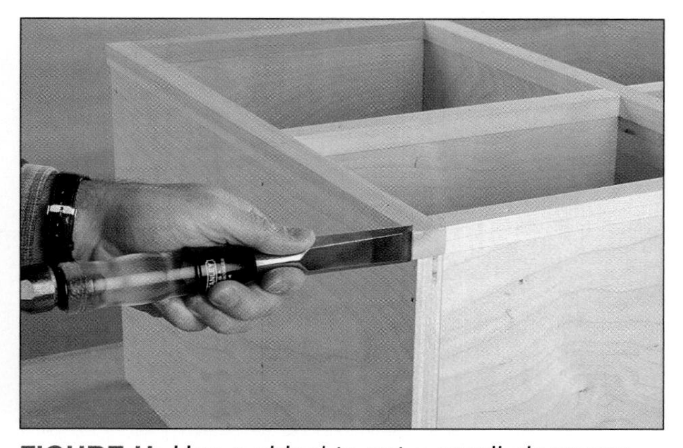

FIGURE H: Use a chisel to cut a small clearance notch in the top corner of each vertical facing strip. Cut in toward the case.

10 Rip and crosscut the edge strips, Parts F1 and F2, for the top.

11 Apply glue to them, clamp them to the top, and nail the parts together.

12 Cut the poplar facing pieces, Parts G1 through G4, to size.

13 Apply the facing to the front of the sides, top, bottom, middle shelf and partitions. Start with the case sides, then apply the facing to the horizontal parts and finally to the partitions. Note that the strips overhang the plywood panels by 1/16 in. on each edge except for the case top, which has a 1/8-in. overhang.

14 Use a chisel to cut the notch at the top outside corners of the facing strips *(see Figure H).*

15 Place the top panel upside down on the work surface, and invert the case assembly over it.

16 Bore pilot holes, and screw the top to the assembly.

17 Complete the case by cutting the back, Part H, to size and nailing it in place.

18 Rip and crosscut the pieces of poplar and plywood for the toe kick assembly, Parts I through K.

19 Clamp the assembly together, and join the parts with glue and 6d finish nails.

20 Clamp the toe kick assembly to the bottom.

21 Bore and countersink pilot holes through the cleats into the bottom.

22 Screw the cleats to the bottom.

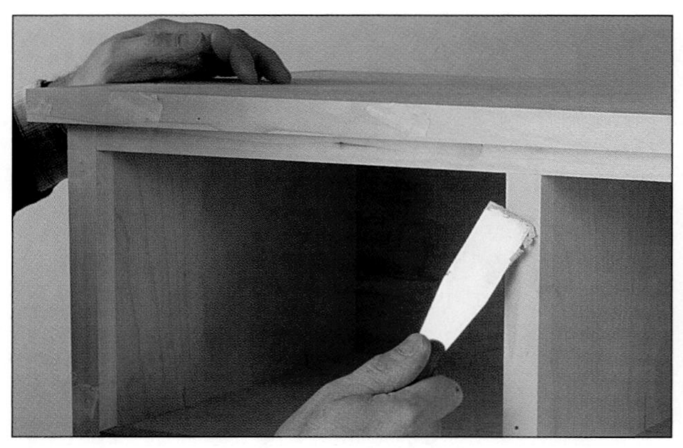

FIGURE I: Use a putty knife to press drying filler into the nail holes. Slightly mound the filler, and let it harden before sanding.

FIGURE J: Sand the surfaces carefully using a random-orbit block sander. This tool is small enough to fit into the compartments.

FIGURE K: Put a small, crisp bevel on the facing and edge strips with a sanding block that you move perpendicular to the strip's edge.

FIGURE L: A small-diameter roller is used to apply the primer and top coat. The square end of the roller allows it to paint into corners.

Finishing

1 Set the heads of all finish nails below the surface.

2 Fill the holes with a wood filler *(see Figure I)*. Mound the filler slightly over each hole since it shrinks when it dries.

3 Sand the cabinet, inside and out, with 120-, 150- and 180-grit sandpaper *(see Figure J)*. Remove all sanding dust before moving to the next finer grit of sandpaper.

4 Carefully ease all sharp edges with a sanding block *(see Figure K)*. Move the sanding block perpendicular to the wood's edge to achieve a crisp bevel.

5 Remove all sanding dust by vacuuming and using a tack cloth before applying the primer.

6 Use a small-diameter, smooth-surface paintroller to apply a coat of latex primer to all cabinet surfaces *(see Figure L)*. Note that the long-handled roller used here has one end that is somewhat shaggy. This allows you to apply paint right to the corner.

7 When the primer is dry, sand it lightly with 220-grit sandpaper.

8 Finish the project by applying two coats of latex semigloss paint for an attractive finish.

Installing Home Security

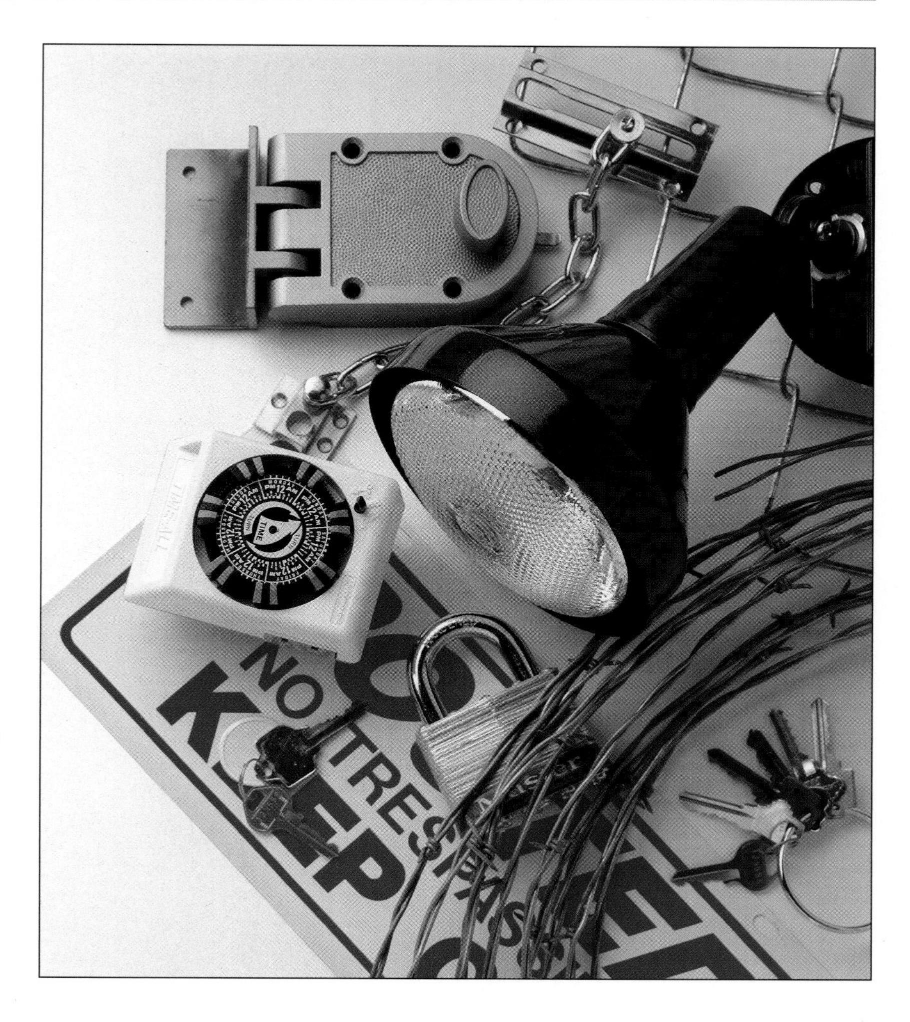

If you've ever had your home broken into, you know the sense of violation it leaves behind. Homes are private spaces, and they're supposed to be free of public intrusion. And quite often these days, they're work spaces too, crammed with computers, scanners and fax machines alongside all the expensive equipment and pricey toys that define our leisure hours—just the things that turn your home into a target.

So how do you safeguard against intrusion and theft? There are two common approaches—high-tech and low-tech—and both start with the understanding that if a professional thief really wants to get in, it will probably happen. So protection in the real world is a matter of degree.

The reasonable assumption is that most thieves are opportunists, and they'll choose an easier target over a more difficult one most of the time. If it takes too long to gain entry, if breaking in will create too much noise, and if once inside, there are still further obstacles such as motion sensors and alarms, then the chances of getting caught increase. If you do enough of the right things to raise the risk factor sufficiently, the balance begins to shift in your favor.

If your home's been burglarized before, or if it houses an expensive hobby collection or lots of business equipment, you may wish to consider a professionally designed, installed and monitored security system. These systems have sensor and alarm components wired to a control panel that is connected to a central monitoring station via phone lines. Any breach of the system is investigated immediately, either by the security company or the police.

The problem with these systems is that they're costly to install and you'll pay a hefty monthly monitoring fee as well. So unless you feel especially vulnerable, a simpler, low-tech approach may work about as well. The strategies we've chosen are based on affordable products you'll find at your neighborhood home center or building supplies dealer. And, you can install them yourself in a weekend or two with common household tools.

Begin with Door Locks

If your locksets appear worn or have a sloppy feel to them, your first step is to replace them with new, quality units. The difference between a flimsy lockset and a good one is only about $10, and this first line of defense is no place to skimp.

Each exterior door should also have a deadbolt lock. There are two types: double-cylinder locks, keyed on both sides, and single-cylinder versions where the inside is controlled by a latch knob. Double-cylinder locks offer better security, but many codes don't allow them because they can interfere with escape in case of fire. We installed a deadbolt that's keyed on just one side (Weiser No. NDC-9470, about $17, Weiser Lock, 6700 Weiser Lock Dr., Tucson, AZ 85746).

Most new doors are steel-clad with either wood or steel-clad edge trims. If yours has a wooden edge, install the lock as you would in a wooden door. If it has steel-clad edges, like ours, look for a plastic insert covering the lock area.

If you have a sliding door, its lock is the weak link in your perimeter defense. The simplest and most effective solution is to bar the sliding half against the jamb of the stationary half. To do the job, we chose a Patio Security Bar (about $26, Master Lock, 2600 N. 32nd St., Milwaukee, WI 53210).

The bar fits openings between 29⅛ in. and 43⅞ in. Just lift the handle and extend the bar until it fits the frame, then lock the pin in place and lower the lever *(see Figure A).*

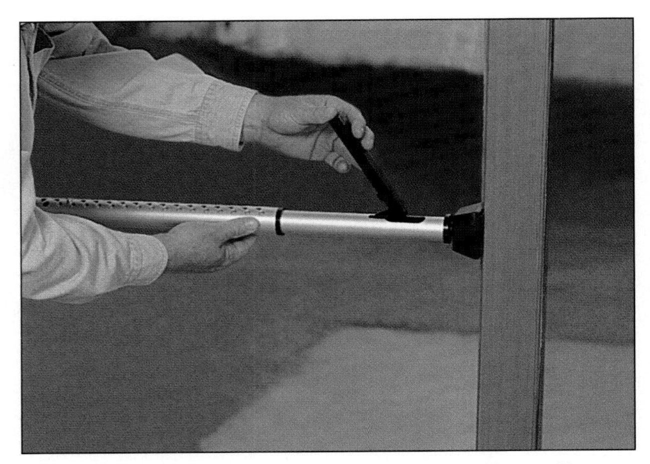

FIGURE A: To install the Patio Security Bar, adjust its length to fit between the sash and the jamb, then lower the tension lever.

Installing a Deadbolt Lock

1 A plastic insert has score marks, corresponding to different bolt plate sizes. Pick the size you need and cut through the insert with a utility knife *(see Figure B).* Then break out that part of the insert.

FIGURE B: For a steel-clad door with steel-clad edges, find the plastic cover at the lock area. Follow the scored marks to cut away the cover.

2 If you find a steel plate with machined screwholes, use these holes to dictate the placement of the bolt. With the paper template supplied by the manufacturer, mark the centers of the bolt and the lock.

3 Using these marks, bore through the face of the door with a 1½-in. holesaw *(see Figure C).* When the pilot bit just pokes through, finish the hole from the opposite side.

FIGURE C: Use the lock's template to mark the hole center on the door face. Then, cut the hole with a 1½-in. holesaw.

4 Bore into the door's edge with a 1-in.

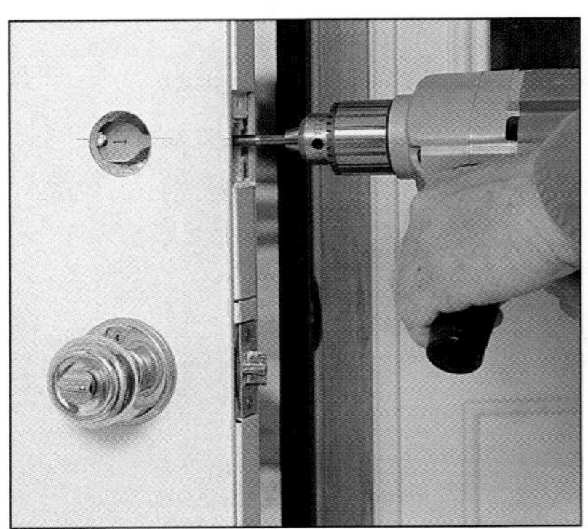

FIGURE D: Bore the hole for the bolt with a 1-in. spade bit. Bore past the far edge of the face hole to provide room for the bolt.

spade bit. To accommodate the throw of the bolt, bore ½ in. past the far edge of the lock hole *(see Figure D).*

5 Insert the bolt into the edge hole and screw the plate in place *(see Figure E).*

6 Slide the latch side of the lock in so it engages the bolt mechanism, and install the cylinder half of the lock. Secure these components with the provided screws.

7 Look for a raised nub on the end of the bolt. Color this nub with lipstick or a crayon *(see Figure F),* close the door and operate the bolt to leave a spot of color on the jamb.

8 With the spot as a center mark, hold the lock's strike plate against the jamb and trace around it with a sharp utility knife.

9 Bore into the jamb with a 1-in. spade bit.

10 Chisel out the strike-plate area to about 1⁄16-in. deep.

11 Secure the plate with the supplied mounting screws.

FIGURE E: Slide the bolt in and screw the plate to the door. Make sure the bolt hub, visible in the face hole, is oriented down.

FIGURE F: Apply lipstick or crayon to the nub on the bolt end. Operate the bolt with the door closed to mark the strike plate location.

Secure Your Windows

Locks for double-hung windows are a problem, because most work only with the window closed. If you'd like a little ventilation, you have to surrender your security. To resolve this problem, we installed a Fortress sash lock (about $8.50, VSI Donner, 12930 Bradley Ave., Sylmar, CA 91342). The lock comes with two plates so the sash can be secured in two positions.

Installing a Window Sash Lock

1 Screw a lock plate to the lower corner of the upper sash with the lower sash closed *(see Figure G).*

2 Insert the lock's bolt through this plate and screw the lock to the top of the lower sash *(see Figure H).*

3 Install a second lock plate several inches above the first, allowing the window to be locked in a slightly open position *(see Figure I).*

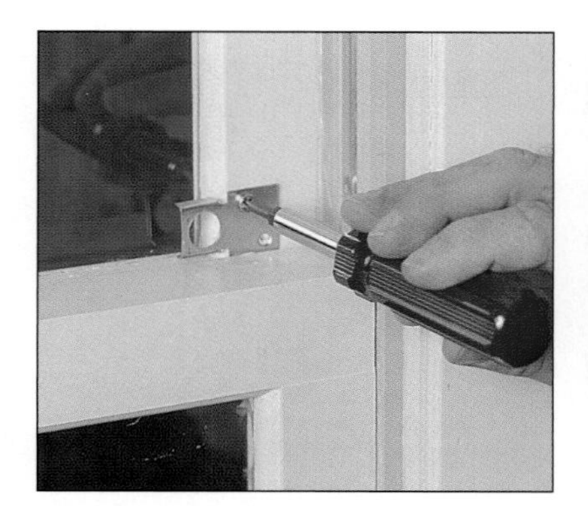

FIGURE G: To install a Fortress window lock, screw the first lock plate to the lower corner of the upper sash with the window closed.

FIGURE H: Position the lock on the lower sash so its bolt fits the hole in the first lock plate. Then, screw the lock in place.

FIGURE I: Finally, install the second lock plate a few inches higher. This allows the window to be locked in an open position.

Motion Detection

We opted for a PIR motion-sensing Wireless Ceiling Alarm (about $30, Homewatch Security, Lamson Home Products, 25701 Science Park Dr., Cleveland, OH 44122). This device senses heat movement within 20 ft. and sounds a 100 dB alarm. It's activated by a compact remote that fits a key ring.

FIGURE J: To install a ceiling-mounted motion sensor, use plastic anchors to screw the metal bracket to the ceiling.

Attaching a Wireless Motion Detector

1 Pick a spot above a traffic area and screw the metal mounting bracket to the ceiling *(see Figure J).*

2 Attach the sensor and secure it with two side screws *(see Figure K).*

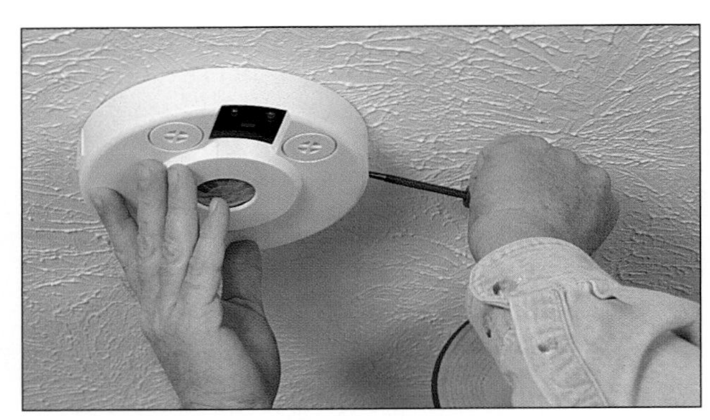

FIGURE K: Attach the sensor to the bracket, using two side screws. This unit is powered by two C batteries.

Security Lighting

Breaking and entering is a lot easier in the dark, so a motion-sensing floodlight is a good idea. Because a new light will require some new wiring and an indoor switch, check with local code authorities before starting the work.

While the light we installed (Intelectron, No. BC9000R) is no longer in production, you still may find it on store shelves. Similar units are available, such as the model SL541ZWA (about $25) made by Heath Zenith, P.O. Box 90004, Bowling Green, KY 42102.

We wired the new light to an underutilized 15-amp circuit, through an existing ceiling fixture. We then installed it on the gable end of the house, about 12 ft. off the ground.

Installing a Security Light

1 Shut off power to the circuit and drop the existing ceiling fixture.

2 Go into the attic and bore down into the interior wall that will house the new switch *(see Figure L).*

3 Feed two 14/2 w/g cables into the wall cavity and staple both cables to a nearby joist.

FIGURE L: After choosing the switch location on an interior wall, move to the attic and bore a hole through the wall's top plate.

FIGURE M: Feed two cables into the wall—one to bring power and one to feed the new fixture. Anchor the cables with staples.

FIGURE N: At the switch, join white wires and secure black wires to switch terminals. Attach grounds to the grounding screw.

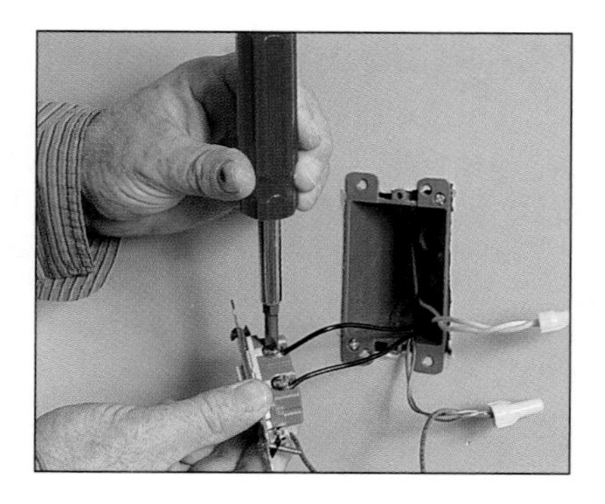

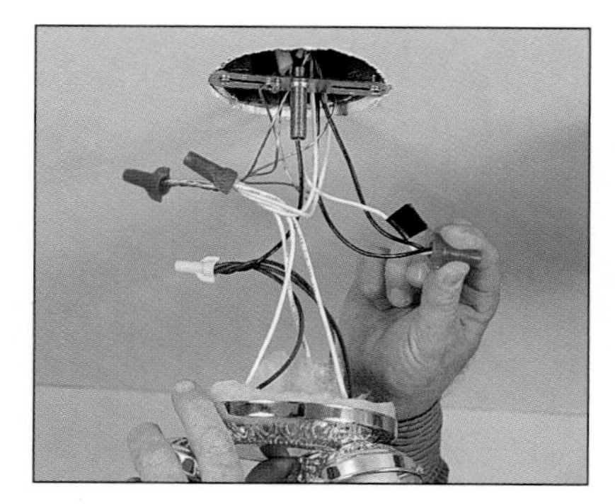

FIGURE O: Join the black wire from new switch to coded white and black hot wires. Join all white wires with twist connector.

FIGURE P: At the new fixture, join black and white switch wires to black and white leads. Attach ground wire to grounding strap.

FIGURE Q: Attach the fixture to the box with its fastening nut, caulk all seams and aim the sensor out and slightly downward.

4 Run one cable to the existing ceiling box and the other to the security light location. Staple both cables every 4 ft. and within 8 in. of their respective box connections *(see Figure M).*

5 To install the switch, hold a plastic cut-in type box against the wall, 44 in. off the floor, trace around it and cut the opening.

6 Pull the two cables from the wall cavity into the back of the box.

7 Strip the sheathing from the ends of the cables.

8 Press the box into its wall opening and secure it by tightening its screws.

9 Join white wires in one twist connector, ground wires to a grounding pigtail attached to the switch's grounding screw and black wires to the switch terminals *(see Figure N).*

10 Our ceiling fixture contained two white wires, two black wires and two bare ground wires. One of the white wires is coded with black tape to indicate that it's the hot feed to the existing switch. The remaining white wire is the neutral side of the lighting circuit. To bring power to the new switch, remove the twist connector from the coded white wire and black hot wire and join the black wire from the switch to these wires. Then join the white wire from the switch to the existing white wires *(see Figure O).*

11 Using a 1-in. spade bit, bore into the gable end of the attic at the location of the new light, and feed the cable from the new switch through this opening.

12 Thread a short, galvanized nipple into the back of an exterior weather box and pull the cable into the box through this nipple.

13 Fasten the box to the siding with galvanized deck screws.

14 Strip the sheathing from the exposed cable.

15 Join the black switch loop wire with the black fixture leads and the white wire with the white fixture leads. Attach the ground wire to the fixture's mounting strap, or as local codes dictate *(see Figure P).*

16 After installing the fixture, aim the sensor and caulk between the box and the siding and the box and fixture cover *(see Figure Q).*

Text and Photos by Merle Henkenius

Inside Insulation

If you've looked into insulation lately, you've seen the avalanche of facts and figures. You've heard the industry bickering over health, safety and performance claims with marginal proof presented as carved in stone. For an industry so well established, with products so well tested, this may be surprising. After all, except for lightweight urethane foam, the materials used in our homes today were commonplace 20 years ago, and most of them were around long before that. Even the latest installation enhancements are now over 10 years old.

So what's the bottom line? All insulations work, but some work better than others in specific situations. This has always been true. What's made insulation more complicated today is the fact that building homes has become more complicated. While it's agreed that sealing our homes against the elements brings greater comfort and lower fuel bills, the last 20 years of progress have brought new worries like sick-house syndrome and structural rot. Insulations and vapor barriers can cause these problems—but they can also help solve them. To understand which insulation materials and related products work best in various parts of your home, it pays to begin with a look at how heat moves and the ways to help make it stay put.

Insulation Basics

Wherever you find two areas of different temperature, you'll find heat transfer—from hot to cold. In the case of your home, this heat transfer takes place in three primary ways. First, it takes place by heat conduction through the material your house is made of. Second, it moves by convection, that is, by currents of air within the house, attic and wall cavities. And finally, heat leaves your home by air leaking in or out through cracks, gaps and holes in the structure.

Insulating materials control the transfer of heat by slowing down the rate at which it moves through your home's envelope. R-value is the rating used to indicate a material's ability to resist the flow of heat. The ½-in. drywall that usually covers interior walls, for example, has an R-value of about .4, while typical fiberglass insulation rates at about 3.5 per inch—just over R-12 for a 3½-in.-deep wall cavity.

Filling your walls and attic with insulation, though, is only part of the solution. Air infiltration through openings in the house envelope can account for 30 to 40 percent of a typical home's heat loss. Traditionally, builders used building paper, or tar-paper, to stem the tide. Today, plastic vapor-permeable house wraps are typical. And, don't forget caulk and spray foam for handling gaps around windows and doors, and around holes at electrical and plumbing entrances.

Handling Moisture

A typical family of four will generate 2.5 gal. of water vapor every 24 hours through bathing, cooking, washing clothes—and simply breathing. This moisture, like heat, moves from warm to cold areas—and, the colder it is outdoors, the greater the outward pressure.

This natural migration can pose real problems for today's moisture-vul-

nerable building products. Water vapor moves through ceilings fairly easily, without doing much damage, but walls are closed and are more likely to trap moisture. Even a little dampness can degrade the R-value of most wall insulation. And when the vapor condenses, it can rot structural members.

To prevent moisture from entering the wall cavities, builders use vapor barriers. Kraft paper or foil-faced insulation slows moisture penetration and is a good choice in moderate climates that have few really cold days. In northern states, however, vapor-impermeable polyethylene film is installed on all exterior walls, just under the drywall.

Except in a few very cold locations, plastic vapor barriers are not usually recommended on ceilings. Sealing a house this completely not only traps water vapors inside, but also harmful gases that may be present, like formaldehyde, radon and carbon monoxide.

Insulation Types

Fiberglass and rock wool are considered mineral insulations because they're made, at least in part, from sand or rock. Since the '50s fiberglass has become the more popular of the two and, in fact, it now dominates the entire insulation market.

Fiberglass comes in two forms—batts and loose-fill. Batts are cut to fit 16- or 24-in. framing spaces and come in 70-ft. rolls or cut packages of 48- or 93-in. lengths. Batts are available with faced and unfaced sides. When faced, kraft paper or a thin aluminum foil is glued to one side to serve as a vapor barrier. Common batt thicknesses are 3½, 5½, 6 and 8 in.

Fiberglass batts are made in various R-values. Greater resistance can be designed in by altering the density or fiber length. A 3½-in. fiberglass batt, for example, is available in R-11, R-13 and R-15 versions, with commensurate pricing.

A more recent form is encapsulated fiberglass—batts in sealed plastic bags. It's not as itchy to install and it reduces airborne fibers. However, a good number of batts need to be trimmed in some fashion, and cutting encapsulated insulation is more difficult and just as itchy. It also costs 15 to 20 percent more.

Loose-fill fiberglass is designed to be blown in place. The traditional place for loose-fill is in attics and closed-wall retrofits, but fiberglass is also blown into open walls in new construction and remodels. Known as blown-in blankets, the fiberglass is blown into stud cavities through holes in a netting or polyethylene film stapled to the studs. Blown-in blankets have three times the density of batts and are better at slowing convection. R-values are around 4.1 per inch, or 14.3 in a 3½-in. wall. Blankets require more work and equipment, however, and you'll pay 35 to 45 percent more than for batts.

Cellulose is made from ground-up newsprint and other recycled paper products, with boric acid added as a fire retardant. By its nature, cellulose has greater density than fiberglass, which makes it more resistant to air infiltration and convection currents. Its R-value is around 3.7 per inch. In our market, it was also the most affordable.

Like fiberglass, cellulose

Insulation Cost Table

Insulation type	Location	R-Value	Materials Cost	Installed Price
Fiberglass, batt	walls	R-13	$250	$450
Fiberglass, batt	attic	R-30	$450	$800
Fiberglass, blown	attic	R-30	$430	$490
Fiberglass, blown	walls*	R-14	NA	$1000
Blown-in blanket	walls	R-14.3	NA	$740
Cellulose, blown	attic	R-30	$350	$450
Cellulose, blown	walls*	R-14	$175	$700
Cellulose, wet	walls	R-14	NA	$650
Polyurethane spray foam	walls	R-24.8	NA	$1450
Icynene spray foam	walls	R-13	NA	$1200
Expanded polystyrene	varies	R-4	$4**	NA
Extruded polystyrene	varies	R-5	$7**	NA
Polyisocyanurate	varies	R-7	$15**	NA

* Existing wall retrofit. **Cost for 4 × 8 panel. NA = does not apply.

The dollar amounts shown are based on a sampling from a Midwestern market—prices may vary considerably in other locales. The figures are based on a 1000-sq.-ft. home, 1000-sq.-ft. attic, 3½-in. stud cavities and a flat ceiling. Wall area is rounded off at 1000 sq. ft. Foam board R-values are based on 1-in. thickness.

has been used for decades in blown-in attic and retrofit wall situations. And like fiberglass, it can now be blown into open cavities in new construction. In this application, the cellulose contains an adhesive. When water is added, it can be shot into the cavities with a machine. Any excess thickness or overspray is scraped away. Wet application cellulose makes a very tight wall—tighter than fiberglass or loose-fill cellulose—with an R-value near 14 in a 3½-in. cavity. It, too, is labor intensive and more expensive.

Insulating rigid-foam boards are made of either expanded polystyrene (the white beadboard that looks like coffee cups), extruded polystyrene (colored blue, pink, yellow or green depending on manufacturer) and polyisocyanurate, a spongy, less brittle foam with stiff paper or foil backing. Foam boards range between R-4 and R-7 per inch and are available in many different sizes with 4 × 8 ft. and 4 × 9 ft. panels being two of the most popular. Beadboard is also available in stud-space widths and is commonly used between firring strips on basement walls. Foam board intended as exterior sheathing is usually coated with a paper, plastic or foil film, to increase nailing strength.

The main selling point for polyurethane spray-in-place foam is that it seals cavities almost completely, stopping convection and infiltration. That's a real plus, and polyurethane also boasts the highest R-value of any insulation, delivering around R-7 per inch.

On the downside, it's expensive, though prices vary widely by location. With the growing emphasis on tight houses, though, polyurethane is gaining ground, and builders frequently use it in combination with other materials to reduce costs while delivering high R-values.

In the past five years, a modified, lightweight urethane foam has gained a foothold in the United States. Under the trade name Icynene, this foam is applied like polyurethane, but water is used as a propellant. Icynene remains soft and pillowy when set, which means it can expand and contract with the structure. It's been thoroughly tested,

TYPICAL INSULATION USAGE

A conventional insulation strategy includes attic, wall and basement insulation, plus a vapor barrier on the inside walls and a vapor-permeable house wrap on the outside. Spray-in-place foam may not require a vapor barrier. Some localities may not require house wrap if foam board sheathing seams and nail holes are sealed.

and emits no detectable vapors after 30 days, so it's a good choice for those with chemical allergies. Its R-value is only 3.6 per inch, but again, its air sealing capacity is its primary appeal. Spray foams are vapor impermeable, cutting down on the need for vapor barriers. Icynene also comes in a pour-in formulation, designed for walls in existing homes.

Retrofit Strategies

If you're thinking of adding insulation to your existing home, but can't do it all at once, prioritize from the top down. Heat rises and it only makes sense that you'd lose more heat at the ceiling and upper walls than at waist level. If your house is uninsulated, 6 in. of new attic insulation will pay for itself in as little as a year or two—and you'll improve comfort. After that, the law of diminishing returns comes into play. Another 6 in. may take seven to eight years to pay for itself, and not noticeably improve comfort.

Moreover, the effectiveness of the first 6 in. means you're now losing relatively more heat through your walls than before, so your next investment should be insulation in the walls, including the rim joists. After this, it's time to focus on the windows, always the weak link in heat loss and now more vulnerable to condensation. Of course, replacing your windows can be very expensive, which can mean a long payback period. Caulking around the windows' exterior trim, replacing the weatherstripping around the sashes and adding heat-shrink plastic in the winter are low-cost alternatives that can make a big difference. Once you've taken care of your windows, it's time to return your attention to the ceiling. After adding insulation there, focus on your basement or crawlspace walls.

As for insulation choices, pour-in Icynene is an attractive option for a drafty old house in the Rust Belt or atop a hill on the High Plains. Otherwise, most homeowners opt for traditional materials, cellulose or fiberglass, in the walls and attic.

As for attic fiberglass, blown-in is more popular than batts for three reasons. First, attics are uncomfortable work spaces, which can promote sloppy installations. They're also filled with structural bracing and wires, which means tailored cuts to avoid gaps that rob R-value and concentrate moisture penetration. And finally, loose-fill makes better contact with the ceiling joists to slow heat loss through the wood.

When blowing fiberglass in an attic, be sure to install vent chutes between the rafters over each soffit vent and use batts between the other rafters where they pass over the exterior walls. This keeps insulation out of the soffit vents and wind from blowing the insulation inward.

If you prefer batt insulation for the whole job, the first layer should be thick enough to fill the joist cavity, and the second layer should be laid perpendicular to the first, with tight edges, so the tops of the joists are well covered.

Studies suggest that penetrating convection currents in attics can rob R-value from fiberglass, but only on very cold days. If you don't have many sub-zero days, it's not a big issue. If you do, cellulose is a better choice.

New-Home Options

If you're insulating a new home, there are two distinct approaches. The first is a conventionally insulated and sealed home—batts or blown-in blankets in the walls and either batts or loose-fill in the attic, plus foamed or stuffed utility holes between floors. This approach usually includes a plastic vapor barrier on the walls and a vapor-permeable house wrap under the siding. It does not usually include a vapor barrier in the ceiling. Water vapor and other gases, like radon and carbon monoxide, are allowed to rise through the insulated ceiling or escape through the usual gaps and cracks. Probably 90 percent of the houses built today are of this type. The problem is that some leak too much, wasting energy, and others don't leak enough, so that noxious gases are trapped inside.

The second option is more extreme. In this case, the goal is to maximize R-value and to eliminate leakage altogether, which means plenty of insulation and sealing every surface and every possible opening. Because the walls and ceilings are vapor sealed, roof-vented furnaces and water heaters aren't used. The same is true for leaky ceiling lights and outlet boxes and, quite often, bath and kitchen fans, wood-burning fireplaces and unsealed sump pits. In this system, stale air is expelled and fresh air is brought into the house through a mechanical ventilation system and associated ductwork. These measures are effective, but the added expense (2 to 5 percent of the cost of the house) goes directly to the issue of affordability.

In new construction, regional building codes dictate minimal standards, but you may well want to exceed them. Your step-up choices will be fiberglass batts with higher R-values, building thicker exterior walls to hold more insulation, blown-in blankets, wet-installation cellulose and, finally, spray foams. All are good options, but each represents a price increase.

You can also mix materials to increase performance and decrease costs. You could use blown-in blankets in the walls and cellulose in the attic. Or just install 1 in. of urethane spray foam in the walls to seal them against air infiltration, and fill the remaining space with fiberglass batts. Finally, you could foam the trouble spots—the rim joists, headers, outside corners—and insulate the rest of the house with batts.

New-home attic insulation is essentially the same as described for retrofits. When installing batt insulation in vaulted ceilings, size the batts to leave at least 1 in. of airflow above them. New roofs usually have continuous soffit and ridge vents, so that top inch becomes the airway. When installing batts in floors, especially over crawlspaces, install them faced side up.

Codes may also require foam board sheathing below grade, against the foundation or the basement walls. When required, it's usually 1 in., but in some cases, material that is 4-in. thick is required. Foam board sheathing may also be required on the footings and walls of crawlspaces and, occasionally, under basement floors and slabs in conjunction with vapor barriers.

Cost and Payback

State energy offices, utility companies and installers can answer the investment/payback question, but each home is different, so general rules are hard to come by.

So how can you know which insulating approach is best for you? Climate matters, utility rates matter and years in your home matter, because it usually takes years to earn back the difference. Common sense also helps. If you spend a lot of time thinking about utility bills, then take a step up when you build or add insulation to your existing home. If your utility costs leave you breathless, then invest as much as you can on insulation and efficient heating equipment. But if your gas company has you on a $40 monthly budget, do the simple things and get on with your life.

Text and Photos by Merle Henkenius
Illustration by George Retseck

Information Sources

Blow-in-Blanket Contractors Association
110 Breeds Hill Rd., Suite 3, 3rd Floor
Hyannis, MA 02601
www.bibca.org

Cellulose Insulation
Manufacturers Association
136 S. Keowee St.
Dayton, OH 45402
www.cellulose.org

Insulation Contractors
Association of America
1321 Duke St., Suite 303
Alexandria, VA 22314
www.insulate.org

The National Insulation Association
99 Canal Center Plaza, Suite 222
Alexandria, VA 22314
www.insulation.org

North American Insulation
Manufacturers Association
44 Canal Center Plaza, Suite 310
Alexandria, VA 22314
www.naima.org

Polyisocyanurate Insulation
Manufacturers Association
1331 F St. N.W., Suite 975
Washington, D.C. 20004
www.pima.org

Window Replacement

Windows provide more than a view of the neighborhood: They're really important parts of your home's protective envelope. Designed to let in light and air, they also hold the elements at bay. When a window begins to fail—to warp, rot, leak or simply refuse to open and close—the costs in terms of both energy loss and aggravation quickly add up.

While new windows are no one's favorite way to spend money, without them you'll eventually spend the equivalent in higher heating and cooling costs—and you'll still have bad windows. The good news is that there are more choices in do-it-yourself window replacement today than ever before.

The simplest replacements are 100 percent vinyl window kits. Here, you remove all but the external frame—the window casing—and have a complete replacement insert custom made to fit the opening. The replacement consists of an insulated vinyl casing with double-hung vinyl windows already installed. To reduce costs, you can order the package with single-hung windows, in which only the bottom window opens. Vinyl windows are also available with low-E glass for reduced heat infiltration.

The next step up is a wooden sash kit, which often comes with vinyl-clad external components. The vinyl exterior is rotproof and never needs painting, while the interior components resemble a standard window. These sash kits do not come preassembled, however. Instead of sliding in a complete insert, you fit new window case components and then install the sashes. You'll pay a little more, but you'll get a tight window with a conventional appearance.

Casement window kits are also available for a little more money. All the usual add-ons, like mullion grilles, screens and low-E glass, are available with these replacements. But while mullion grilles are external on double-hung windows, they are often sandwiched between the glazed panes of casement windows.

Pricing varies with window size and color,

and the type of glass. Assuming a window in the 28 × 54-in. category, prices will range between $103 and $260, depending on features. The lower priced of these are either all plastic or all wood, with double-glazed, insulated glass.

While replacement kits are quick and easy, they can cost as much as a complete window and do nothing for rotted or warped casings and subsills. If your window casings have begun to fail, or if you've never liked the type of window you have, the best approach is a complete window replacement.

We chose a vinyl-clad, double-hung, wood-frame window made by Andersen (model No. 1611859, about $260, Andersen Windows Inc., 100 Fourth Ave., North Bayport, MN 55003). This window has a nailing flange instead of brick molding (exterior trim). We prefer a nailing flange for retrofits because the new brick molding rarely fits the siding. We used 4-in.-wide cedar to trim the outside of our new window.

Installing a Replacement Window

1 To determine the size of the window you'll need, first remove the interior trim with a small pry bar *(see Figure A)*.

2 Measure the old window and the rough wall opening and find a close replacement. A new window of the same general size as the old will often be slightly shorter, owing to a different sill design.

3 With the replacement

FIGURE A: Use a small pry bar to remove the window trim from the inside. Proceed slowly and carefully if you plan to reuse the trim.

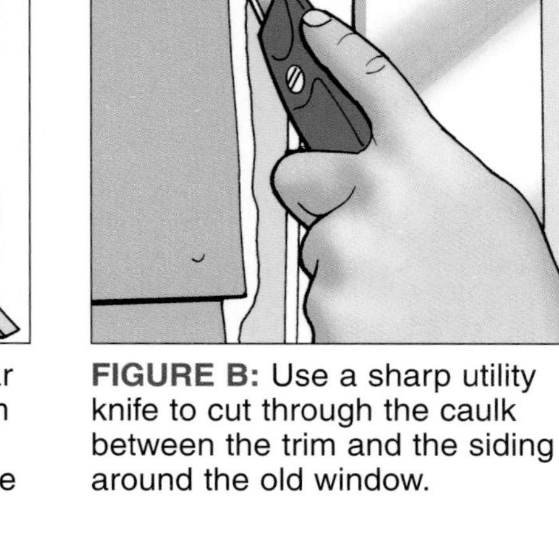

FIGURE B: Use a sharp utility knife to cut through the caulk between the trim and the siding around the old window.

FIGURE C: Pry out all the nails along the outer edge of the brick molding by driving a cat's paw under the nail heads.

FIGURE D: To free the window, tap it loose from the inside with a hammer and block of wood. Strike the casing in several spots.

FIGURE E: As the window loosens, have a helper support it from the outside. When it's free, grip it on both sides and lift it out.

FIGURE F: To remove the old tin drip cap, slide a reciprocating saw blade between the tin and sheathing and cut through the nails.

FIGURE G: To make room for the new window's nailing flange, cut away about 1½ in. of siding from the opening's perimeter.

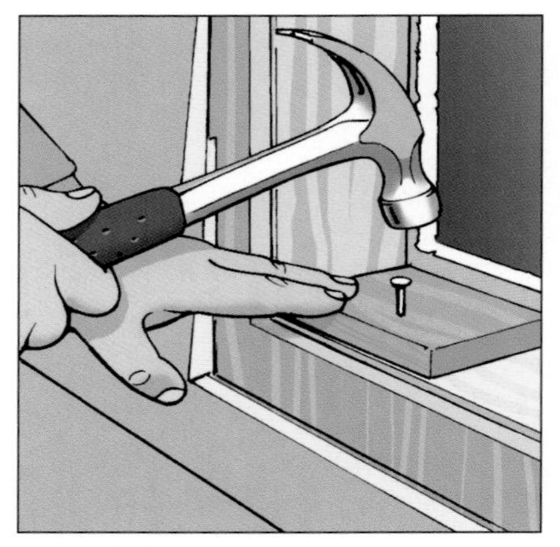

FIGURE H: To raise the new window slightly, nail ¾-in.-thick blocks into the lower corners of the framed opening.

FIGURE I: Set the new window in place for a test fit. Press the flange tightly against the sheathing and center the window.

window on site, go outside and cut through the caulk bead between the old brick mold and the siding *(see Figure B).*

4 Use a cat's paw to pull the casing nails along the outer edge of the old brick molding *(see Figure C).*

5 In most cases, removing the casing nails won't immediately free the window. With a helper outside to steady it, go back inside and loosen the window by tapping with a hammer and block of wood *(see Figure D).* When it breaks free, grip the window along each side and lift it out *(see Figure E).*

6 If the old window had a drip cap along its top, remove it with a reciprocating saw fitted with a hacksaw blade. Just slip the blade between the metal cap and the wall's sheathing and slice through the nails that hold it in place *(see Figure F).*

7 In many cases, you'll also need to cut some of the siding from the edge of the window opening, especially if the original window wasn't centered. This cut is not for the trim boards, which you'll make later, but just to clear a space for the nailing flange. With a circular saw set to a depth of about 1 in., cut out about 1½ in. around the opening *(see Figure G).*

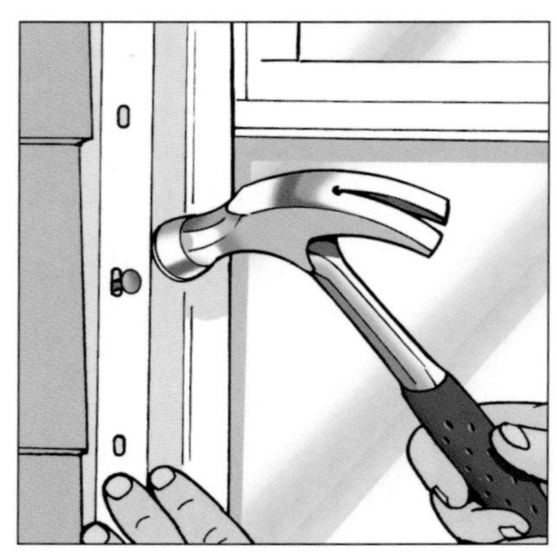

FIGURE J: Use roofing nails to temporarily hold the flange against the wall sheathing. Don't drive the nails completely in.

FIGURE K: Use a 4-ft. level to plumb the window. Shim the casing if needed, then mark the flange position on the sheathing.

8 To center the shorter new window, nail ¾-in. blocks into the lower corners of the framed opening *(see Figure H).*

9 Lift the new window into the opening for a test fit *(see Figure I).* Press the nailing flange firmly against the sheathing, centering the window from side to side as best you can, and tack it in place on both sides with 1¼-in. roofing nails *(see Figure J).* These nails have large, ⅜-in. heads that won't pull through the slots in the nailing flange. Remember that this is just a test fit, so don't sink the nails in all the way.

10 With the window in place, check that it's centered and plumb—a 4-ft. level works best here *(see Figure K).*

11 Framed window openings are seldom level, so you may need to shim the bottom on one side. The best place to shim is between the casing and the spacer blocks you nailed into the corners.

12 When you have the window centered and plumbed, mark the sheathing around the nailing flange for future reference.

13 This is also a good time to mark the siding cut for the 4-in.-wide trim. Just hold a length of trim against the window and mark the siding along the outside of the board on all sides *(see Figure L)*. Because windows expand on hot days, allow ¼ in. between the trim board and the window when marking these cuts.

14 Remove the window to cut out the excess siding along these lines.

15 Before installing the window permanently, slide a new drip cap under the siding along the top. Nail it through the siding with galvanized nails.

16 With the drip cap in place, reinstall the new window, carefully aligning it with the marks from the test fit, and nail it in place. Place a nail about every 8 in. around the nailing flange.

17 Trim the window with the 4-in. stock. Start with the top board, then the bottom and finally the sides. Use 8d galvanized nails to fasten the trim *(see Figure M)*.

18 Carefully apply a bead of paintable silicone caulk to every seam, including the joints between the siding and trim, and the trim and window *(see Figure N)*. Be sure to caulk the seam between the siding and the drip cap, as this joint can trap water and ruin the siding.

19 Finish by priming and painting the trim.

20 Because newer window casings with their vinyl sash guides can spread apart at the center—and because spreading sides can allow the windows to fall out—manufacturers now recommend shimming the casings on each side of the opening. This can be tricky, because if you over-shim, you'll interfere with the operation of the window. Nailing the shims in place through the sashes can also pose problems.

To keep the windows from spreading during shipment, each window has a nylon ribbon that holds the casings together. This band can serve as a guide when shimming the casings. Just slide tapered shims between the casing and the framed opening, one pair on each side. When the shims con-

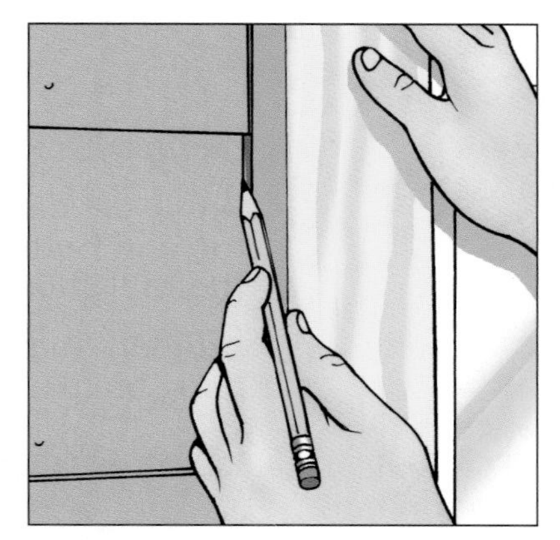

FIGURE L: Lay the trim against the window and mark the siding to be cut away. Leave a ¼-in. gap between the trim and the window.

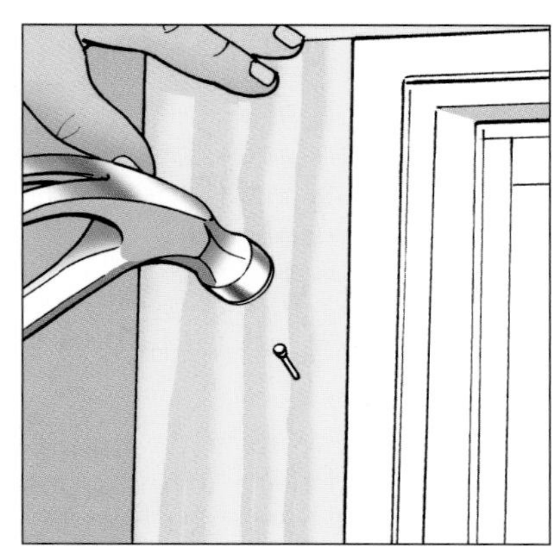

FIGURE M: Install the top and bottom trim boards first, then the sides. Secure the trim with 8d galvanized finishing nails.

FIGURE N: Caulk joints between the trim, siding and window with a high-quality, paintable silicone caulk. Then, prime and paint trim.

FIGURE O: To shim between the casing and side framing, use tapered wooden shims with a dab of construction adhesive on each.

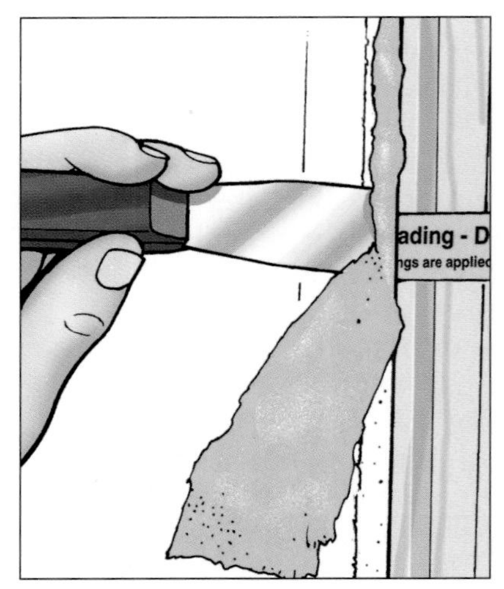

FIGURE P: Use a putty knife to loosely pack fiberglass insulation between the window casing and the wall's framed opening.

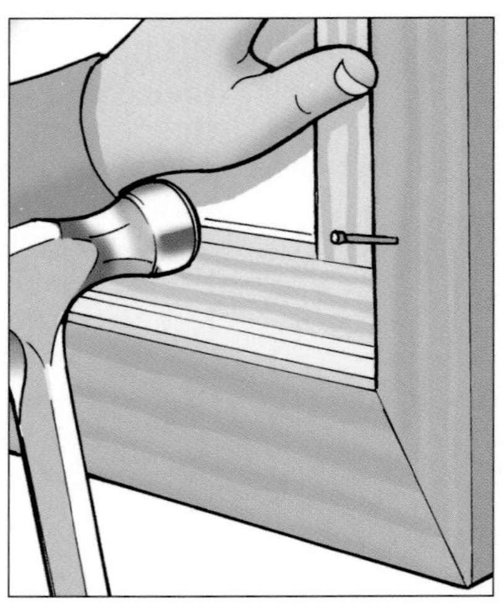

FIGURE Q: Finish by trimming the window casing on the inside wall. Secure the mitered trim with 3d finishing nails. Set nails and fill.

tact both surfaces, but just before they loosen the nylon ribbon, you'll have the right amount of shim *(see Figure O)*.

21 Cut the shims to size and slide them back in place with a dab of construction adhesive on each side.

22 You'll also need to insulate between the window and the framed opening. The easiest approach is to pack fiberglass insulation into the joint *(see Figure P)*. Don't overpack the joint—insulation needs air pockets to work. Another approach is to partially fill the gap with expandable foam. Again, don't overdo it. Use a minimal expansion foam and stop before it expands past the inside edge of the window casing.

23 To complete the job, install the trim on the inside. Start by mitering a length of trim ¼ in. longer than the top header casing.

24 Nail this piece in with ⅛ in. of casing showing on each side, then do the same for the bottom of the window. If you're using a hardwood trim, you'll need to bore pilot holes for the nails to keep the wood from splitting. Use 3d finishing nails, placing one about every 12 in.

25 Finish by mitering and nailing in the side pieces *(see Figure Q)*.

26 Set the nails and fill the holes.

Widening the Opening

If you want a wider window, you'll need to reframe the window opening. There are several ways to approach this work, but here's an easy, low-tech method that nearly anyone can manage.

1 The first step is to cut out the wallboard around the opening. You'll need plenty of working space, so cut well beyond the actual opening.

2 Remove the insulation from the stud spaces.

3 Use a reciprocating saw with a hacksaw blade to cut through the nails that hold the header to the king studs *(see Figure R)*.

4 Cut the nails that secure the studs to the top and bottom plates and those that secure the wall sheathing to the studs.

5 Before removing these structural members, build a small temporary wall to use as

Framing an Opening

Window headers are supported by a two-stud configuration at each side. The outside or king stud is a full-length stud that's nailed to the top and bottom plates. The inside or jack stud is shorter and supports the header and the weight of the roof.

The length of the jack stud is determined by the desired height of the window. The height of a new window on a retrofit usually matches the height of the surrounding windows. In our case, the tops of the other windows are 11 in. down from the ceiling. Allowing 3 in. for the double plate at the top of the wall, we have about 8 in. of header space to fill. And because our window would be less than 4 ft. wide, 2 × 8 header stock would provide enough support. A wider opening may require a larger header, so check with your local building inspector or code authority before you start the job.

a ceiling support. A three-stud, 26- to 36-in.-wide support will usually suffice.

6 Set it in place about 18 in. in front of the window opening and shim it underneath until it makes firm contact with the ceiling *(see Figure S).*

7 Remove the structural members you've cut free.

8 When framing a new opening, make it 1 in. larger than the new window in both directions *(see* Framing an Opening, *previous page).*

9 Cut the jack studs to length and nail them in place against the king studs, one on each side *(see Figure T).*

10 To build the header, cut two 2 × 8s to fit between the king studs and rest on the jack studs.

11 Cut a piece of ½-in. plywood to sandwich between the 2 × 8s.

12 Apply construction adhesive to the first 2 × 8 and set the plywood into this adhesive. Apply adhesive to the plywood, place the second 2 × 8 on top and nail the assembly together with 12d nails driven from both sides.

13 Set the header on the jack studs and nail through the king studs *(see Figure U).*

14 Frame in the bottom of the window opening and cut away any excess siding and sheathing.

By Merle Henkenius
Illustrations by George Retseck

FIGURE R: Use a reciprocating saw with a hacksaw blade to cut nails securing the old header. Do the same for the studs and sheathing.

FIGURE S: Build a temporary support wall to take the load when you remove the old header. Shim it to make contact with the ceiling.

FIGURE T: Nail the jack studs in place against the king studs on each side. Window height determines the length of the jack studs.

FIGURE U: Set the new header atop the jack studs and nail it in place through the king studs. Then, remove the temporary support wall.

Replacing Roof Shingles

Because your roof protects your home, everything inside is threatened when the shingles begin to fail. Having the roof replaced, though, is a labor-intensive job that can cost several thousand dollars—no matter how well-equipped a roofer is, roofing still goes down one shingle at a time.

If writing the check at the end of the job is a problem, a do-it-yourself roof is often a reasonable alternative. While roofing is always a big job, usually taking the homeowner a week or more, there's really nothing very complicated about it. While material costs will vary, a 40% to 50% savings is not out of line. If you shop carefully, you can buy all you need for less than $50 per square (100 sq. ft. of roof area).

Reroofing Options

The simplest approach is to add a new layer of shingles over the old—most municipal codes allow up to three layers. However, more than two layers are impractical for several reasons.

FIGURE A: Use a potato fork to peel asphalt shingles and felt from the roof. Pull or drive in remaining nails and sweep the roof clean.

To begin, each new layer adds thousands of pounds, which can threaten weaker roofs. Secondly, new shingles over old never lay completely flat and can appear lumpy. And finally, new shingles won't last as long when laid over degraded shingles—a third layer can shorten the life span of the new roof from 25 to 15 years.

If your existing roof must be removed, or you simply want to do the job right, first inspect the type and condition of the roof sheathing. From the attic, check to see if you have solid sheathing or spaced sheathing. If you find solid sheathing, the job will be easier and less expensive. But if you find horizontal boards spaced 1 to 3 in. apart and covered with cedar shingles, the price—and aggravation—goes up substantially.

After tearing off all the shingles, you'll need to either put back new cedar shingles, which will cost about $150 a square, or cover the old sheathing with plywood or oriented-strand board, and then finish with asphalt or fiberglass shingles. New sheathing will add about $50 a square to the final price.

Once you start removing your roof, you'll want to work quickly while your home is vulnerable to rain. At the end of the workday cover exposed sheathing with plastic tarps. Shovel the old roofing into trucks or a roll-off Dumpster and pay the city landfill to bury it.

Keep in mind that a more complicated roof may require more material, as extra waste is likely. And while replacing a simple roof may be easy, especially if you have help, a job with several dormers, valleys and hips may justify a professional installation, no matter what the cost. Roofing is also something you should not take on if you're at all uncomfortable with heights. If your roof is steeper than a 4 pitch (4 in. of rise for each 12 in. of run) you should consider hiring a pro.

FIGURE B: Nail gutter apron along the roof eaves. On inside corners, cut the overhang on the first apron to wrap around the corner.

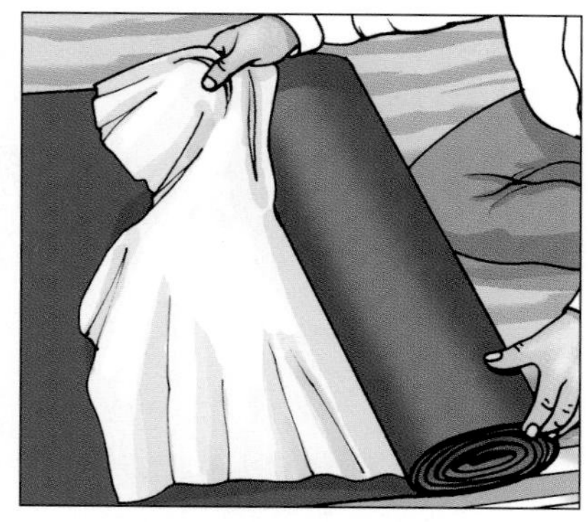

FIGURE C: Roll out the ice-dam membrane, peeling the adhesive's backing film away as you go. Press the membrane to the sheathing.

Getting Started

1 Develop a plan. It's often easier, for example, to reshingle one side of the roof before tearing off the remaining sides. This step-by-step approach offers better weather protection and reduces the amount of plastic you'll need to buy.

2 Before tearing off your roof, lay plastic tarps on the ground beneath the eaves to catch shingle nails and tiny bits of asphalt. Protect shrubs and flower beds by propping 2 × 4s against the house and stapling plastic sheeting over them.

3 Use a square-tipped shovel or a potato fork to tear off the old shingles. Because the tines of a fork can get under the shingles without slamming into the nails, a fork works a little better. Begin by prying up the ridge cap on the side you wish to strip first. Then, just dig in, working side to side and top to bottom *(see Figure A)*.

4 When you accumulate a large pile of shingles, push them into a truck or Dumpster.

5 Pull out any nails with roofing stuck under them and hammer in the clean ones. Then sweep and clean the roof gutters.

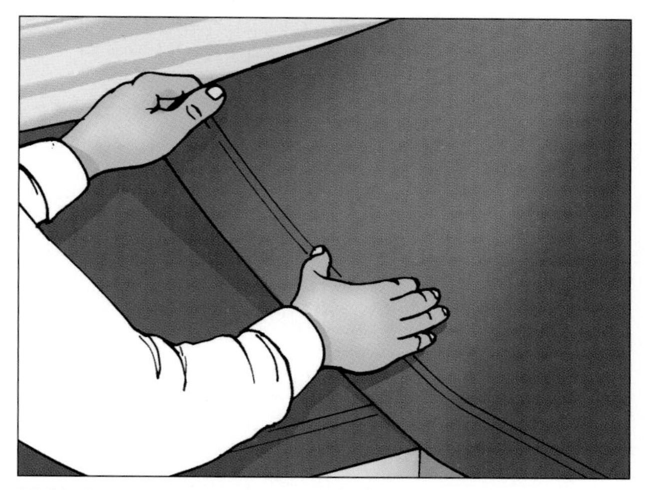

FIGURE D: Allow the ice-dam membrane to pass through the valley in both directions. Then smooth and press to seal adhesive.

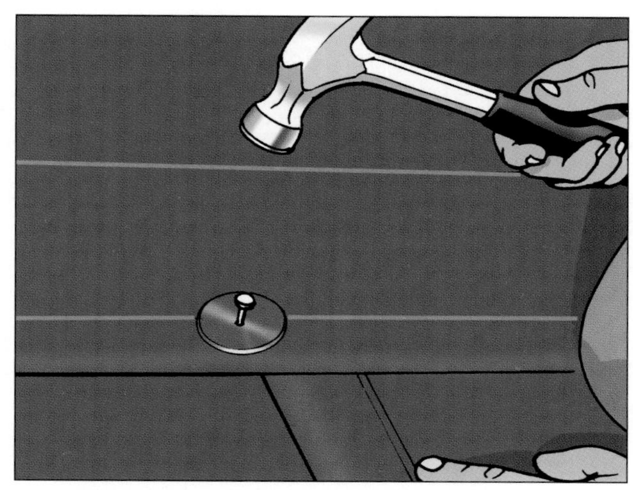

FIGURE E: Install 30-pound roofing felt to the remainder of the roof. Overlap each roll 3 in. Nail in place through metal roofing discs.

Installing Underlayment and Tin

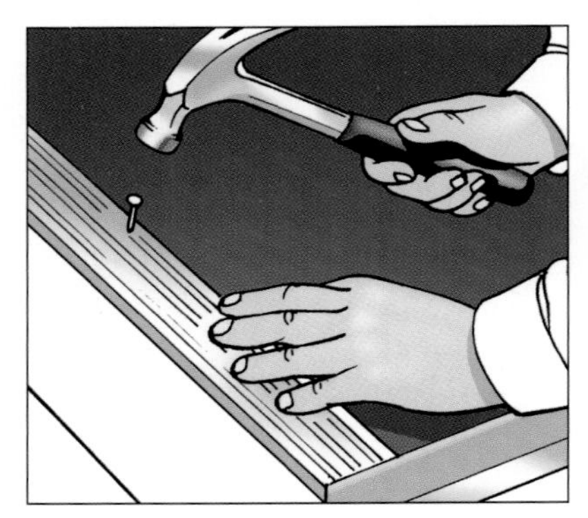

FIGURE F: Nail preformed drip edge over the gable roof edges. Start by overlapping the gutter apron at the eave.

FIGURE G: Trim the end of the valley tin to fit the roof at the lower end of valley. Cut in at 45° from each side, then across the ridge.

1 Nail new gutter apron or drip edge along the eave. Hold it against the edge of the sheathing and place one nail about every 16 in.

2 Outside corners can be cut along the top and wrapped around the corner.

3 At an inside corner, cut the first length slightly longer along the bottom, where it overhangs the sheathing. Then bend that section around the corner. Nail the intersecting length over the bend *(see Figure B).*

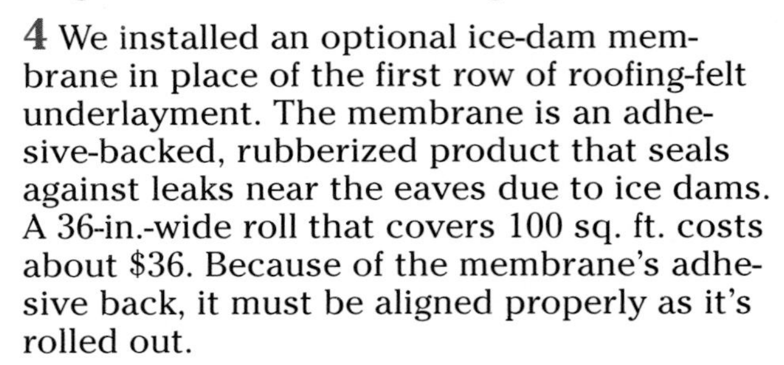

4 We installed an optional ice-dam membrane in place of the first row of roofing-felt underlayment. The membrane is an adhesive-backed, rubberized product that seals against leaks near the eaves due to ice dams. A 36-in.-wide roll that covers 100 sq. ft. costs about $36. Because of the membrane's adhesive back, it must be aligned properly as it's rolled out.

5 Peel the plastic adhesive backing from the first 2 ft. of the roll, then start it along the gutter apron with about a ¼-in. overhang at the edge. Check that the membrane is aligned.

6 Roll it out, peeling away the plastic backing as you go *(see Figure C).* Every few feet, press the membrane against the roof. When the row is done, press down the entire membrane so it sticks to the sheathing.

7 When installing membrane over a valley, run the first roll through the valley at least 3 ft. as measured along the bottom of the roll. Run the intersecting roll completely over the first *(see Figure D).*

8 Cover the rest of the roof with 30-pound roofing felt. Overlap the membrane by at least 3 in. and roll through the valleys. If yours is a hip roof, roll about 2 ft. past the hip and nail it in place. Overlap each row 3 in. and use the marks on the roll to help with alignment. Nail the felt in place through metal roofing discs, placed about every 16 in. *(see Figure E).*

FIGURE H: With the lower end overhanging the gutter apron by ¼ in., nail the valley in place. Nail only the outer edges.

9 With the underlayment finished, nail drip edge along the gable (rake) edge. Start at the bottom, with the drip edge overlapping the gutter apron, and secure with 1¼-in. galvanized roofing nails *(see Figure F).*

10 There are several ways to roof a valley. In some cases, roofers lace the intersecting rows over one another, in other cases, they run through the valley from one side and trim the overlap on the adjacent side. We prefer a formed-tin valley, especially if the roof is not very steep. For a short valley like ours, a 14-in. tin valley is adequate—a longer valley may require 20 in.

11 Begin by trimming the bottom of the length of formed-tin valley to fit the inside corner above the gutter. Cut inward at 45° to within ½ in. of the formed ridge. Then cut across the ridge *(see Figure G)*. This cross cut will deliver the water well away from the apron and underlayment, but not so far that it overshoots the gutter.

12 Lay the trimmed valley in place so that its bottom edge overhangs the apron about ¼ in. and nail it in place *(see Figure H)*. Be sure to nail only the outer edges of the valley.

Installing the Shingles

The most popular layout, or stagger pattern, for standard 12-in.-tab shingles is 6-in. offset, but 4-and-8-in. and 5-and-7-in. patterns are also common. We opted for the usual 6-in. offset because manufacturers provide a notch in the top center of each 12-in. tab—at 6 in.—which makes layout and trimming fast and simple. The disadvantage of using a 6-in. offset is that water travels in streaks down the roof due to the tab slots that line up every other row.

1 If your roof doesn't have a valley like ours, you should measure the overall roof length first. Then calculate your layout so you have the same width shingle on both ends of the roof.

2 Because each shingle is slotted, you'll need to provide a starter row under the first row of shingles. You can buy a solid starter roll, but a reasonable alternative is to remove the tabs from a few shingles and install the remaining pieces as a starter row. Just use a utility knife to score across the back sides of the tabs and break them off *(see Figure I)*.

3 Nail the shingle with the gritty side up and the factory edge down against the gutter apron *(see Figure J)*. Overhang the apron and drip edge by ¼ in.

4 When you come to the valley, apply two beads of plas-

FIGURE I: To make a starter row, trim the tabs from several standard shingles. Cut from the back side to avoid dulling the blade.

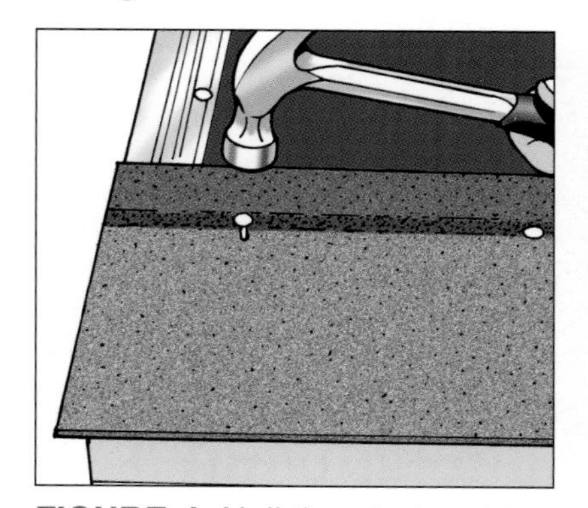

FIGURE J: Nail the starter shingle in place with the factory edge against the apron. Use 1¼-in. galvanized roofing nails.

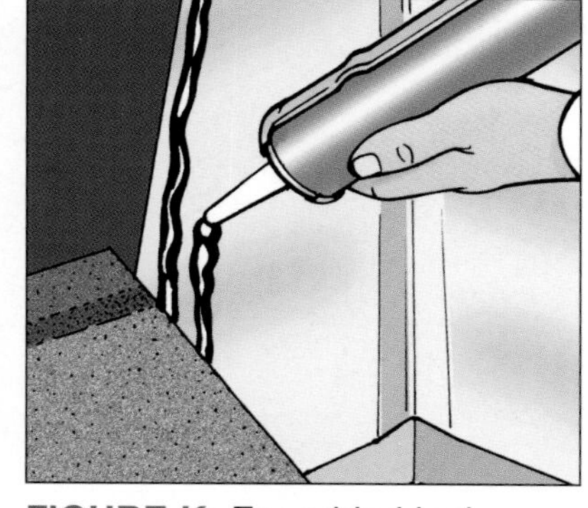

FIGURE K: For added leak protection at the valley, apply two beads of plastic roof cement with a caulk gun. Press each row of shingles into it.

FIGURE L: To provide for the offset on your first row, trim 6 in. from the end of the first shingle. The next row starts with a full shingle.

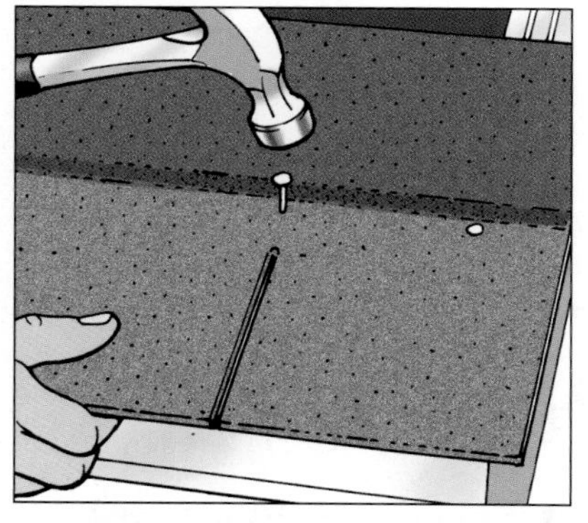

FIGURE M: Nail the first shingle in place with a ¼-in. overhang at the end and bottom. Nail through the inner edge of the drip edge.

FIGURE N: When shingling into the valley, let each shingle run a little long. Then trim all shingles when you're finished.

tic roof cement to the outer edges of the valley *(see Figure K)*. Cut the valley shingles a little long, about to the center of the valley, and avoid nailing through the tin. You'll trim the valley later.

5 To start the first row of shingles, turn the first shingle over and trim 6 in. from the end *(see Figure L)*. If you cut from the back side of the shingle, the blade won't dull as quickly.

6 Nail the trimmed shingle over the starter row. Using 1¼-in. galvanized roofing nails with ⅜-in. heads, nail through the drip edge and 1 in. above each tab slot, for a total of four nails *(see Figure M)*.

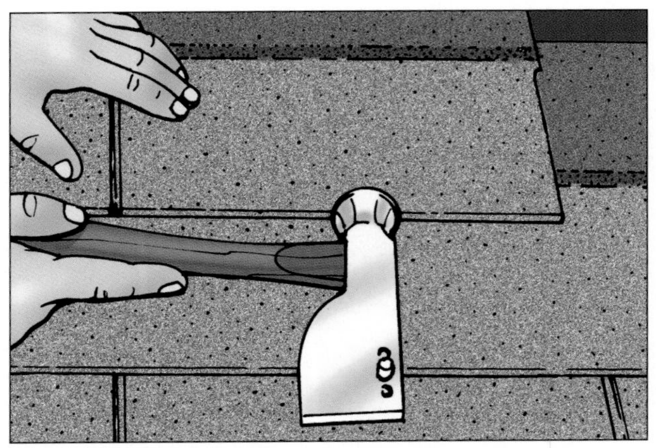

FIGURE O: A roofing ax speeds shingle layout. The peg catches the bottom row and the head marks the placement of the next row.

7 Lay the next shingle in place and nail it above each tab slot. Continue across the roof until you reach the opposite end or a valley *(see Figure N)*.

8 Start the next row with a full-size shingle and continue across the roof. To keep the rows straight, position the bottom edge of the upper shingle along the top of the slots in the lower shingle. (You can also snap chalklines, spaced 5 in. apart, across the length of the roof.)

9 Start the following row with 6 in. of shingle removed and continue up the roof, cutting 6 in. from every other starting shingle.

10 For added speed and accuracy, shingles are best laid with a roofing ax. This tool has a broad, waffled head and a preset peg near the sharp end. By hooking the peg against the lower shingle the head is positioned to align the new shingle *(see Figure O)*.

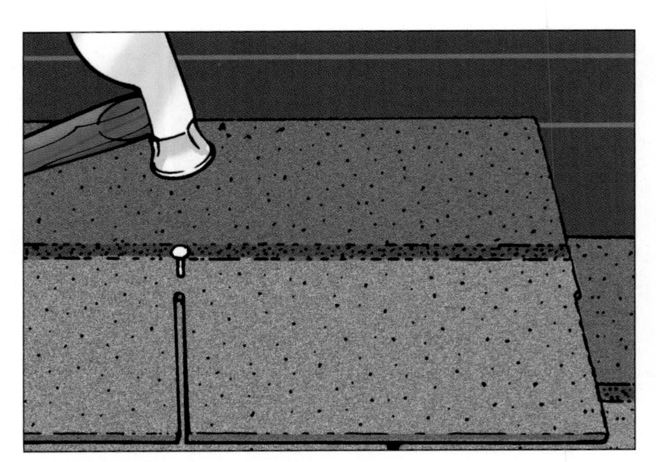

FIGURE P: Nail each shingle four times, once above each tab slot. To save time, install the rows in steps up the starting gable end.

11 It's also faster to step the shingles up along the starting gable edge of the roof instead of installing one complete row at a time *(see Figure P)*.

12 When you reach a plumbing stack, shingle right up to it from below, notching around it as needed.

13 When the shingles are under at least half of the stack flashing, slide the flashing down and nail down its top edge.

14 Apply plastic roof cement around the upper half of the skirt *(see Figure Q)* and trim the next shingle to fit around the stack *(see Figure R)*.

15 To trim the valley, first snap a chalkline along the rough-cut shingles. Because the valley will handle more water at its lower end, flare the line out a little at the bottom. With a 14-in. preformed valley, have a helper hold the line 2 in. away from the center ridge at the top, while you hold it 3 in. away at the bottom *(see Figure S)*.

16 Carefully slice through all layers of shingles along the line *(see Figure T)*.

17 At the peak, we installed a ridge cap made from shingles. Cut a number of shingles into three 12-in. tab-size sections and trim the upper corners from both sides of each piece *(see Figure U)*.

18 When you have a bunch of these tabs cut, nail them over the ridge. Start on one end, placing one nail on each side near the tar strip on the shingle. Then, continue laying at 5-in. intervals until you've covered the entire ridge *(see Figure V)*.

19 Trim the last shingle to fit and nail it with the heads showing.

By Merle Henkenius
Illustrations by
George Retseck

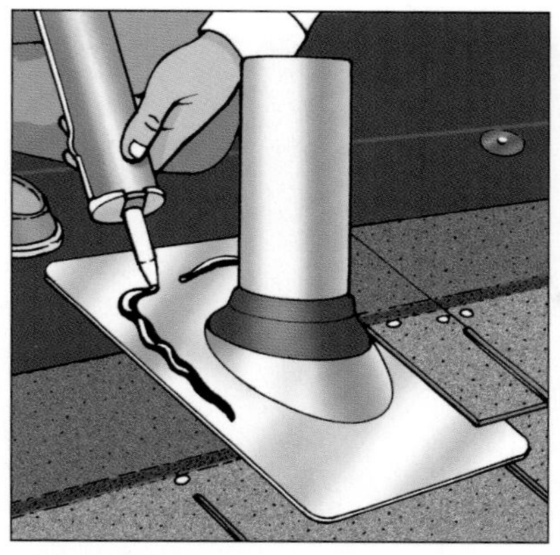

FIGURE Q: To install flashing, shingle the bottom half and place the skirt over the shingles. Apply plastic roof cement to the upper half.

FIGURE R: Cut around the stack at the top and place the trimmed shingle over the flashing. Nail it with nails placed over tab slots.

FIGURE S: To trim a straight line along the valley shingle ends, snap a chalkline, from top to bottom. Make your cut wider at bottom.

FIGURE T: Cut along the chalkline through all of the layers of valley shingles. Use a sharp utility knife and carefully follow the line.

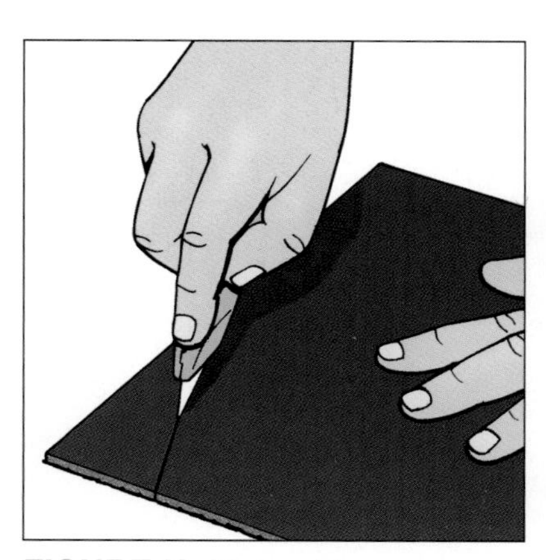

FIGURE U: Make ridge cap shingles by cutting standard shingles into thirds and trimming angled corners at the top edges.

FIGURE V: Nail the caps in place over the ridge in 5-in. intervals. Place one nail on each side. Cut the last tab to fit and nail in place.

Repairing an Icemaker Water Valve

The majority of refrigerators today are equipped with an automatic icemaker. The water valve supplying the icemaker is a key component of the icemaking system, and it should be the first thing you check if the icemaker's performance is erratic or if the icemaker stops working.

When the icemaker calls for ice, its switch closes an electrical circuit and energizes the solenoid-operated water valve. This allows water to flow through the valve and into the ice cube tray. The water is frozen into cubes, and the cubes are dumped into the ice bin.

As time passes, strange things may happen to the refrigerator's icemaking capability. The cubes may be small or there may be a solid chunk of ice instead of individual cubes. It's also possible that the icemaker will stop working. These are all signs of a malfunctioning water valve.

The valve is equipped with a screen on its inlet to remove minerals and sediments in the water supply. Over time, minerals and sediment build up on the screen and restrict flow through the valve, or even block it completely. Minerals that make it through the screen can cause the valve to stick in the open position, overfilling the ice cube tray in the process. This is a common problem in areas with hard water, but it can happen just about anywhere.

Another malfunction that will cause the icemaker to stop working is a break in the solenoid coil winding. This is known as an open coil. The coil winding generates a magnetic field as current passes through it, and this magnetic field opens the plunger valve that controls water flow. A break in the coil winding stops current flow and this prevents the valve from operating.

Test and Inspect

1 The icemaker's valve is easy to inspect and test. First, gently pull the refrigerator away from the wall, and unplug it.

2 Turn off the water supply to the icemaker by closing the shut-off valve in the copper waterline leading to the valve *(see Figure A)*.

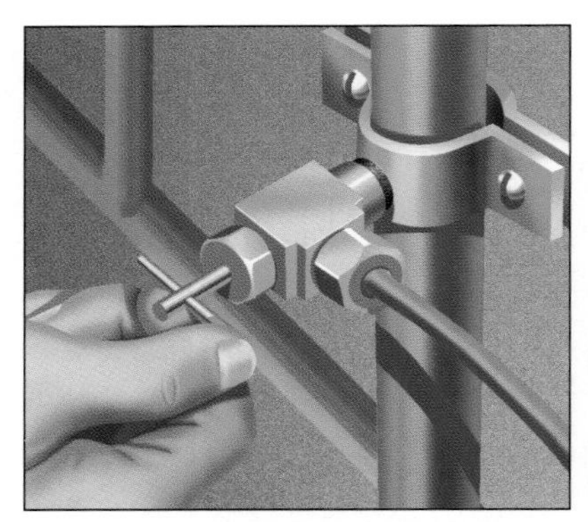

FIGURE A: Begin the repair by unplugging the refrigerator and shutting off the water flow to the icemaker.

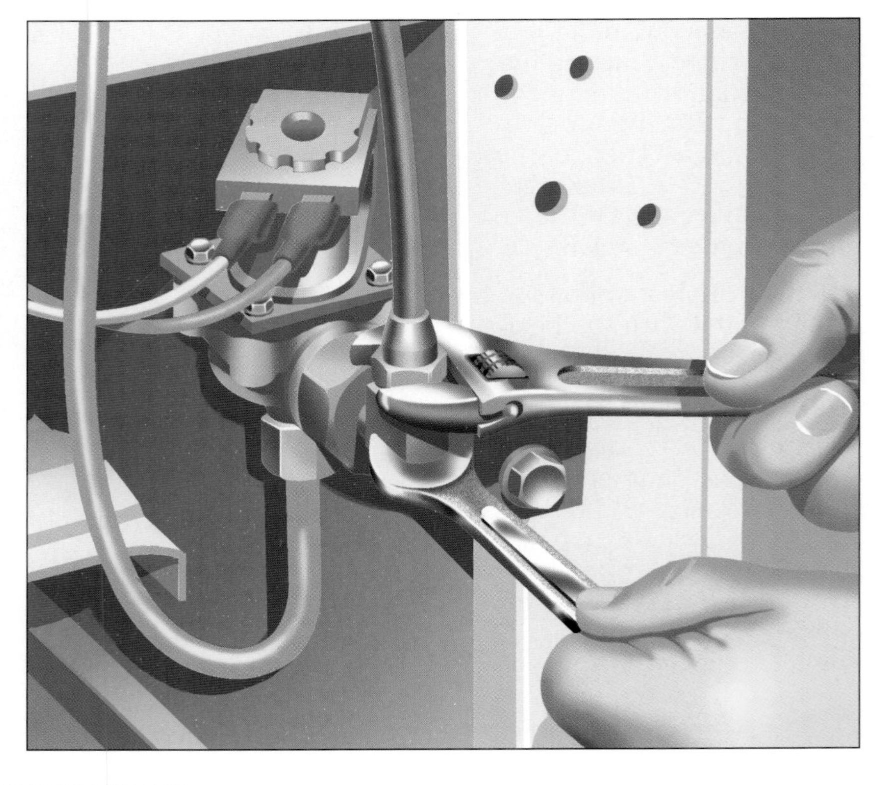

FIGURE B: Loosen the flare nut on the inlet side of the valve with a wrench, using a second wrench to protect the brass fitting.

3 Use a screwdriver or nutdriver to remove the rear lower access panel from the refrigerator's back.

4 Remove the fill tubing from the water valve. Use a wrench to loosen the flare nut on the brass fitting on the inlet side of the valve *(see Figure B)*. Place a container under the valve to catch the small amount of water that will spill from the valve and tubing.

5 Use a screwdriver or a nutdriver to remove the screw holding the valve's mounting bracket to the refrigerator cabinet *(see Figure C)*.

6 Pull the valve out of the compartment and remove the tube on the valve's outlet.

7 Remove the solenoid's electrical contacts *(see Figure D)*.

8 To test the solenoid valve, use a multimeter (also called a volt-ohm meter). Set the meter to the RX-100 scale. Touch the probes to each terminal on the solenoid coil *(see Figure E)*. The meter should read in the range of 200 to 500 ohms.

9 If the meter needle does not move, the coil is bad. You may be able to purchase the coil separately from the valve. If not, you will need to replace the entire valve. An icemaker valve costs about $30 to $50 from the manufacturer or an appliance parts distributor.

10 If the coil tests okay, the inlet filter is probably clogged. To clean the filter, first remove the large brass nut on the inlet side of the valve.

11 Gently pry the screen out with a small screwdriver.

12 Clean the screen using an old toothbrush *(see Figure F)*.

13 Rinse the filter clean, reassemble the valve and install it.

14 Before installing the back panel on the refrigerator cabinet, test run the icemaker. Look for leaks, and tighten any leaky connections. If necessary, use Teflon tape or a similar product to ensure tight connections. Discard the first ice cubes that are produced because they are likely to have sediment in them.

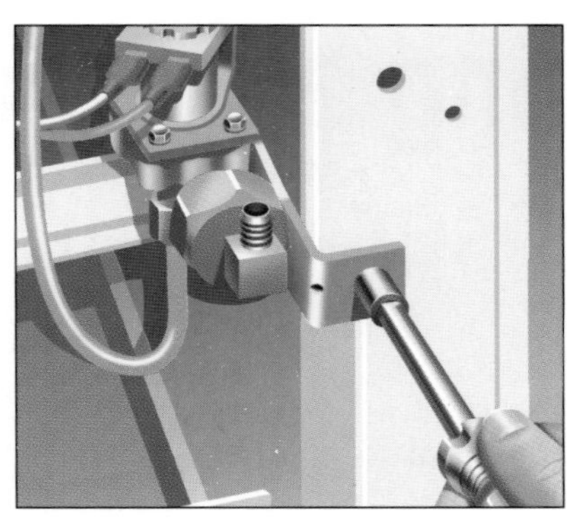

FIGURE C: The icemaker's water valve is attached to the refrigerator cabinet by means of a bracket. Remove the bracket screws.

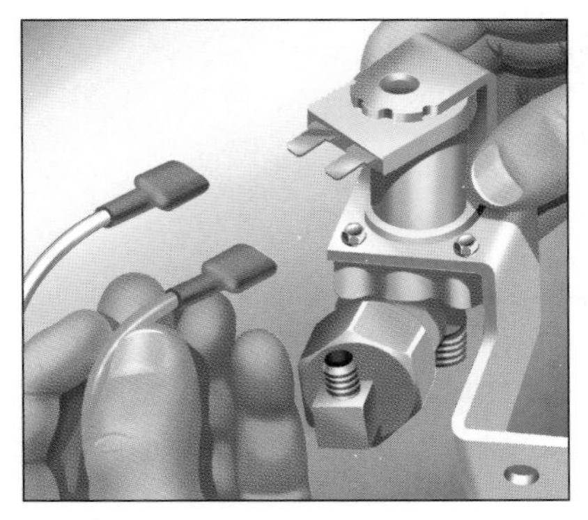

FIGURE D: Two wires supply power to the valve. Gently disconnect the wires to test the solenoid coil's resistance.

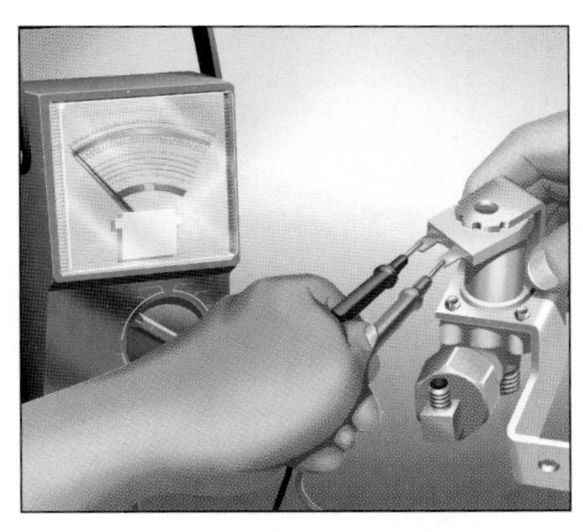

FIGURE E: The multimeter will read 200 to 500 ohms if the coil is good. If the meter's needle does not move, the coil is open.

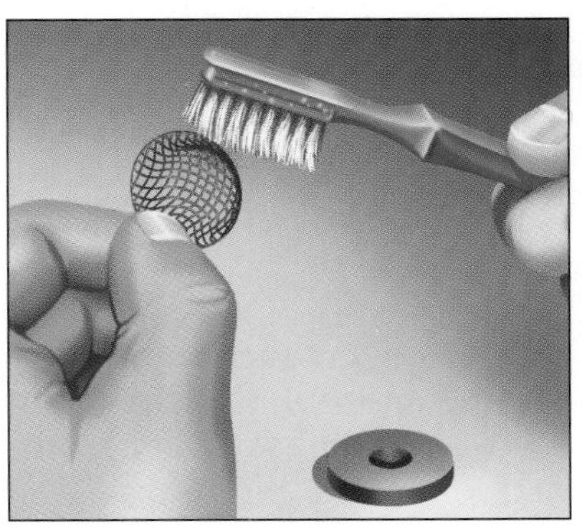

FIGURE F: If the solenoid coil is not malfunctioning, disassemble the valve and gently clean its inlet screen using an old toothbrush.

By Steve Toth
Illustrations by George Retseck

Replacing a Refrigerator Condenser Fan Motor

A condenser fan is used on frost-free refrigerators that have a condenser coil in the bottom of the cabinet, inside the compressor compartment. The fan runs whenever the compressor runs, and it draws cool room air through the front grille, and circulates it through the condenser coils, over the compressor, and back out the front grille into the room. The circulating air helps cool the compressor and the refrigerant in the condenser coils. The air also helps evaporate the water in the drain pan. If the fan stops, the temperature of the refrigerant will rise and the compressor may overheat. Eventually, the food may spoil.

A new condenser fan motor can be obtained from the refrigerator manufacturer or a local appliance parts distributor. When ordering the new motor, supply the company with the make and model number of your refrigerator. Condenser fan motors cost between $40 and $60.

Remove the screws attaching the fan assembly and bracket to the divider in the motor compartment.

Checking the Motor

1 To check the condenser fan motor, first unplug the refrigerator and pull it away from the wall. Remove the rear access cover from the motor compartment.

2 Use a paintbrush to remove the dust from the fan blades, motor and condenser. Then vacuum the area thoroughly.

3 Turn the fan blade to see if it moves freely. If the motor binds or doesn't move at all, replace it.

Replacing the Motor

1 Follow the two motor wires back to the terminal block and remove them using a pair of needle-nose pliers *(see Figure A).*

2 Use either a nut driver or screwdriver to remove the mounting screws that hold the fan assembly and bracket onto the metal divider in the motor compartment *(see Figure above).* Carefully remove the assembly from the motor compartment.

3 Remove the fan blade from the motor by

backing off the thin rectangular nut that secures the fan to the motor *(see Figure B)*.

4 Take the fan blade and rubber washer off the motor shaft.

5 Remove the mounting bracket from the old fan motor and install it on the new motor *(see Figure C)*.

6 Install the rubber washer and the fan blade on the new motor.

7 Tighten the nut, reinstall the fan assembly and attach the motor wires to the terminal block. Check that there are no wires in the way of the blade or motor and that the fan spins freely.

8 Install the rear access cover, then plug the refrigerator back in and test run it.

By Steve Toth
Illustrations by George Retseck

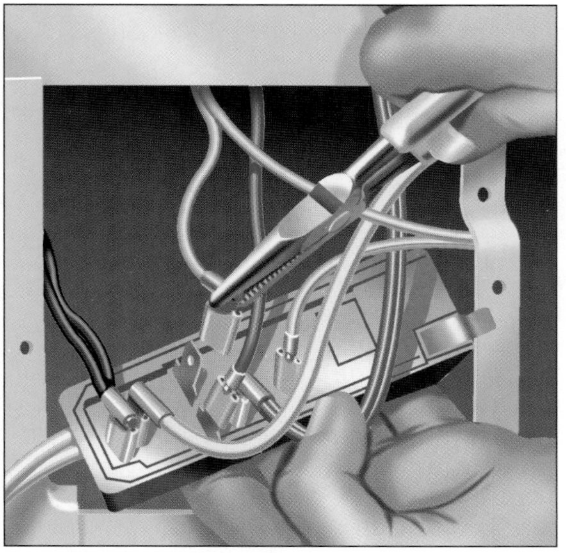

FIGURE A: Trace the motor wires back to the terminal block, and carefully remove them using a pair of needle-nose pliers.

FIGURE B: With the fan and bracket unbolted, remove the thin rectangular nut that holds the fan to the motor.

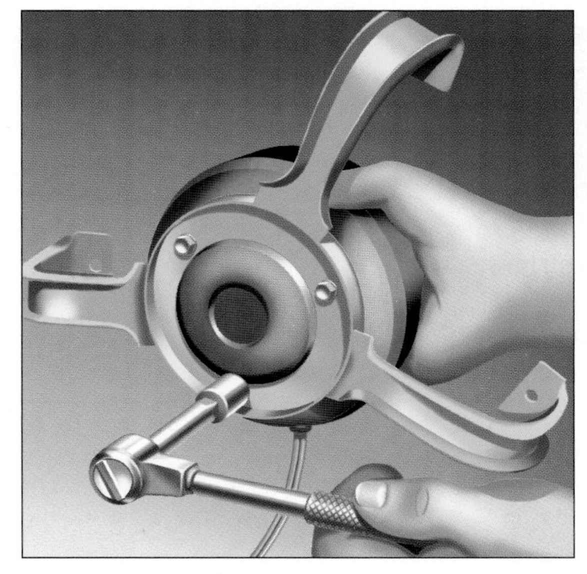

FIGURE C: Use a nut driver or socket wrench to remove the machine screws that attach the fan motor to its bracket.

Replacing a Washing Machine Mixing Valve

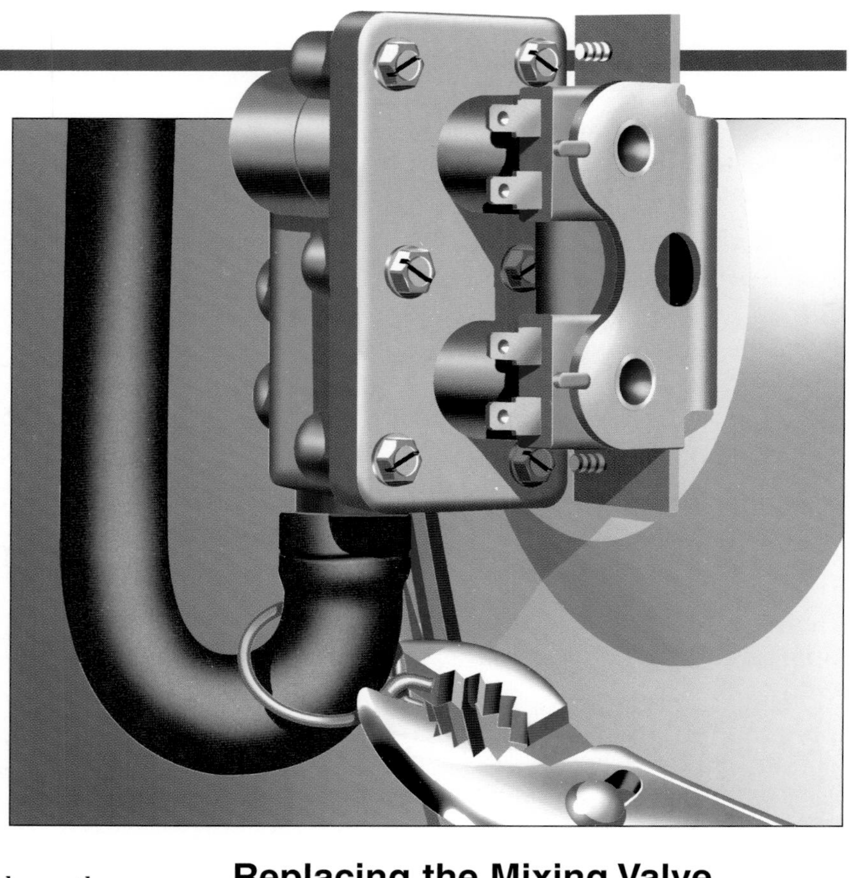

If no water enters your washing machine during the fill cycle, or if only cold water is entering the appliance, there is a strong possibility that the water mixing valve needs to be replaced. Before replacing the valve, however, you should check that the hose screens are clear. There may be screens both where the hoses mount to the valve and where the hoses connect at the faucet. If the screens are clear, then one or both solenoid coils on the valve may be open, or something inside the valve may be preventing it from functioning properly.

We show a water mixing valve on a direct-drive Whirlpool washer. The same procedure would apply to Kenmore machines, and the repair procedure for many other makes of washers is quite similar to what we show here. Regardless of the make, you will also need the model number of the washer to order the replacement part.

Replacing the Mixing Valve

1 Unplug the washer. Then shut off both the hot- and cold-water faucets that supply water to the appliance.

2 Move the washer away from the wall so you can get at the fill hoses on the back of the machine.

3 Using pliers, remove the top hose for the cold-water supply *(see Figure A)*. Use a bucket to catch water that drains out.

4 Wrap a piece of tape around the lower hose for the hot-water supply. The tape will

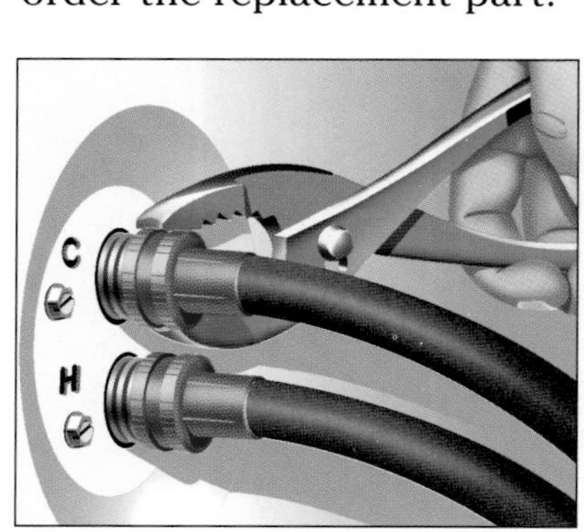

FIGURE A: Remove the water supply hoses using a pair of pliers. Note the position of the hoses. Cold water enters at the top.

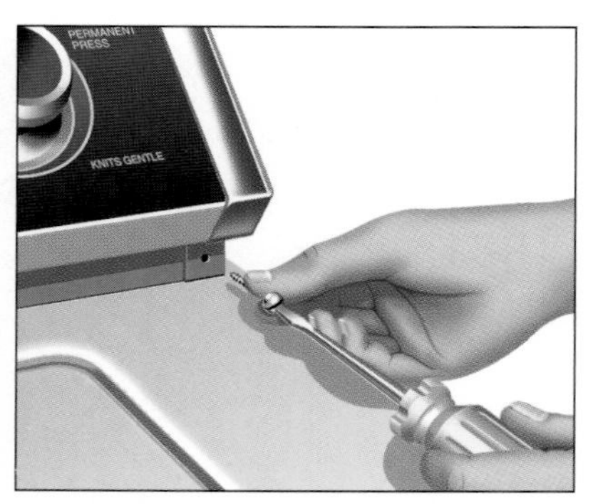

FIGURE B: Before you can pivot back the control panel, you must remove two screws. There is one screw on each endcap.

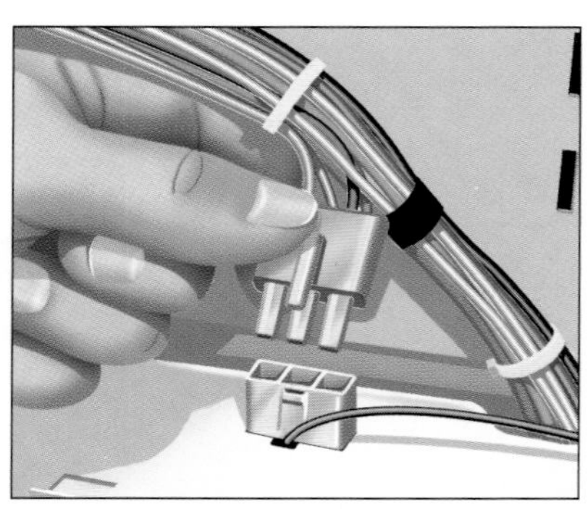

FIGURE C: Pivot back the control panel, and let it rest on its hinges. Disconnect the wire harness by pressing its tabs in.

make it easier to identify it when putting it back on the new valve. Remove the hot-water hose.

5 Remove the machine's cabinet to gain access to the mixing valve, which is mounted on the inside of the rear panel. To begin doing this, use a screwdriver, and remove the screws from the plastic endcaps on the control panel *(see Figure B).*

6 While holding the control panel at each end, gently pull it forward and up. Swing it back so it rests on its hinges.

7 Press the two tabs on the wire-harness connector, and pull straight up on the connector to separate the tabs *(see Figure C).*

8 Remove the two metal clips that hold the rear panel to the cabinet top. Insert a screwdriver into the front edge of the clip. Push down, and move the handle of the screwdriver toward the washer's back *(see Figure D).* This will pry the clip out of the cabinet.

9 Remove the other end of the clip by pulling it straight down out of its slot. Remove the other clip in the same fashion.

10 Hold open the washer's lid, and with your other hand gripping the cabinet at the front, tilt the cabinet forward, and pull it up *(see Figure E).* This will release the cabinet from the tabs on the washer's base.

11 With the cabinet removed you will see the water mixing valve attached to the rear panel. Remove one wire connector at a time, and label it so you will know which coil to attach it to on the new valve.

12 Using pliers, squeeze the spring clamp on the hose and move it back *(see illustration, page 136).* Pull the hose off the valve.

13 Move around to the back of the washer, and use a nutdriver to back out the two screws holding the mixing valve to the rear panel.

14 Pull the valve from the panel, and the removal process is complete *(see Figures F and G).*

15 Install the new valve in the reverse order, and reassemble the appliance.

16 Turn the water on, and check the new valve for leaks to complete the repair.

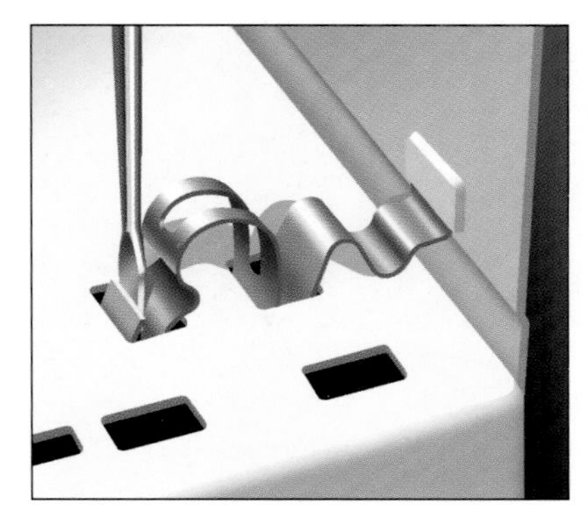

FIGURE D: Insert a screwdriver into the slot at the end of each cabinet clip. Pivot the tool back to remove the clip.

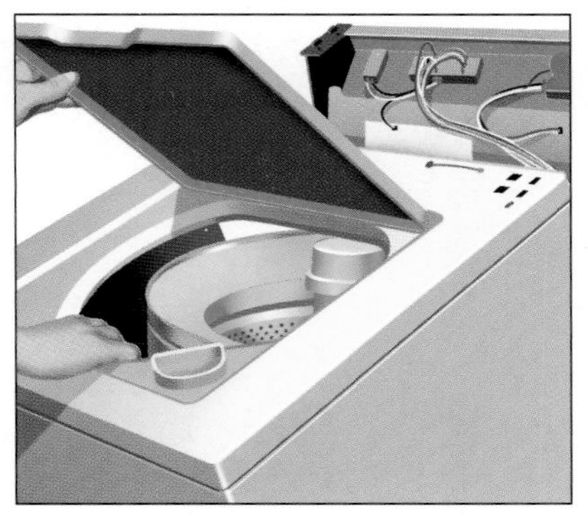

FIGURE E: Open the washer's lid, and while holding it open, grip the cabinet at the front. Pivot the cabinet forward and up.

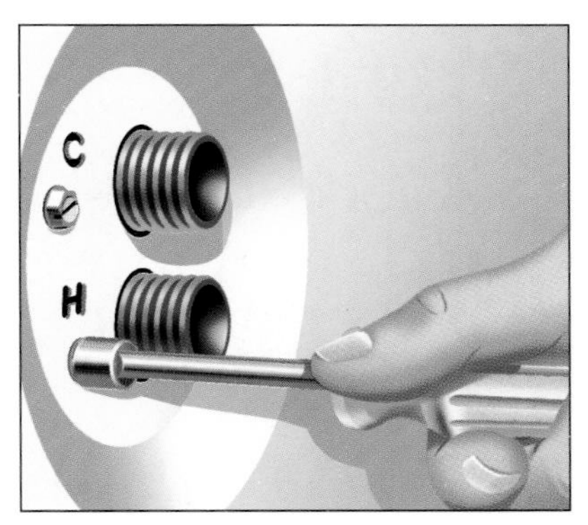

FIGURE F: Use a nutdriver, and remove the hex-head sheetmetal screws that hold the mixing valve to the back panel.

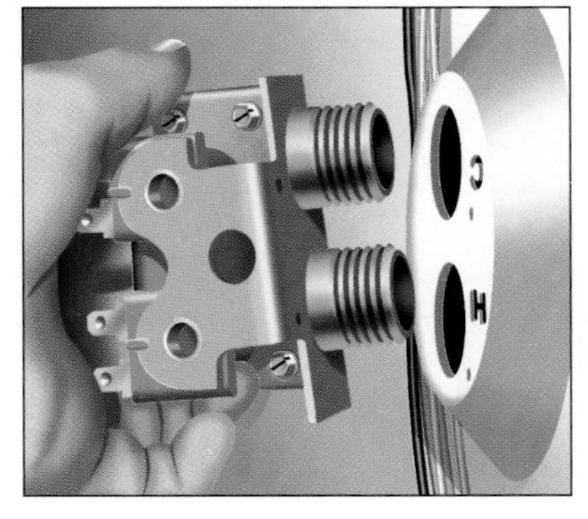

FIGURE G: Move to the inside face of the back panel, and pull the mixing valve's nozzles through their holes.

By Steve Toth
Illustrations by George Retseck

Traditional Picket Fence

Homeowners build fences for a lot of different reasons. Some want total backyard privacy while others can live with a partial screen—enough to peek at the neighbors without feeling on display. And then there are the practical types who lean toward chain link for keeping the kids in and the neighbor's dog out. However, you don't need kids, dogs or something to hide to want a fence. The fact is, a well-designed fence is one of the best ways to add personality to your home. If it also defines your space in a useful way—so much the better.

Of all the fences you might build, the picket fence is one of the most popular. Depending on how it's detailed, a picket fence is as appropriate in a community of split-levels as it is in the backyard of an urban townhouse. The partial screen of a picket fence adds an element of friendly privacy to the landscape—you're not shutting the world out, you're just organizing it.

Building a Fence

Our fence is made up of 1 × 4 pickets screwed to 2 × 4 rails. While the picket/rail assembly is conventional, our post design has several unique features. First, the posts are boxed—¾-in.-thick pine encases a pressure-treated 4 × 4 core. Then, instead of full-length 4 × 4s, ours extend from below the frost line to about 20 in. above grade. This stub post negates the effect of excessive twist common in longer lengths of 4 × 4 stock.

The box post design also makes it easy to notch in the rails and completely enclose the rail ends. At the top, our post has a 4 × 4 core block with a pyramid-shaped upper end. Surrounding the block and recessed into it is a sloped collar that seals out the weather. Decorative trim underneath the collar completes the post top.

This fence is designed for a relatively level site. If your site has minor elevation variations, plan for the tops of the pickets and posts and the top rails to be level. Cut the

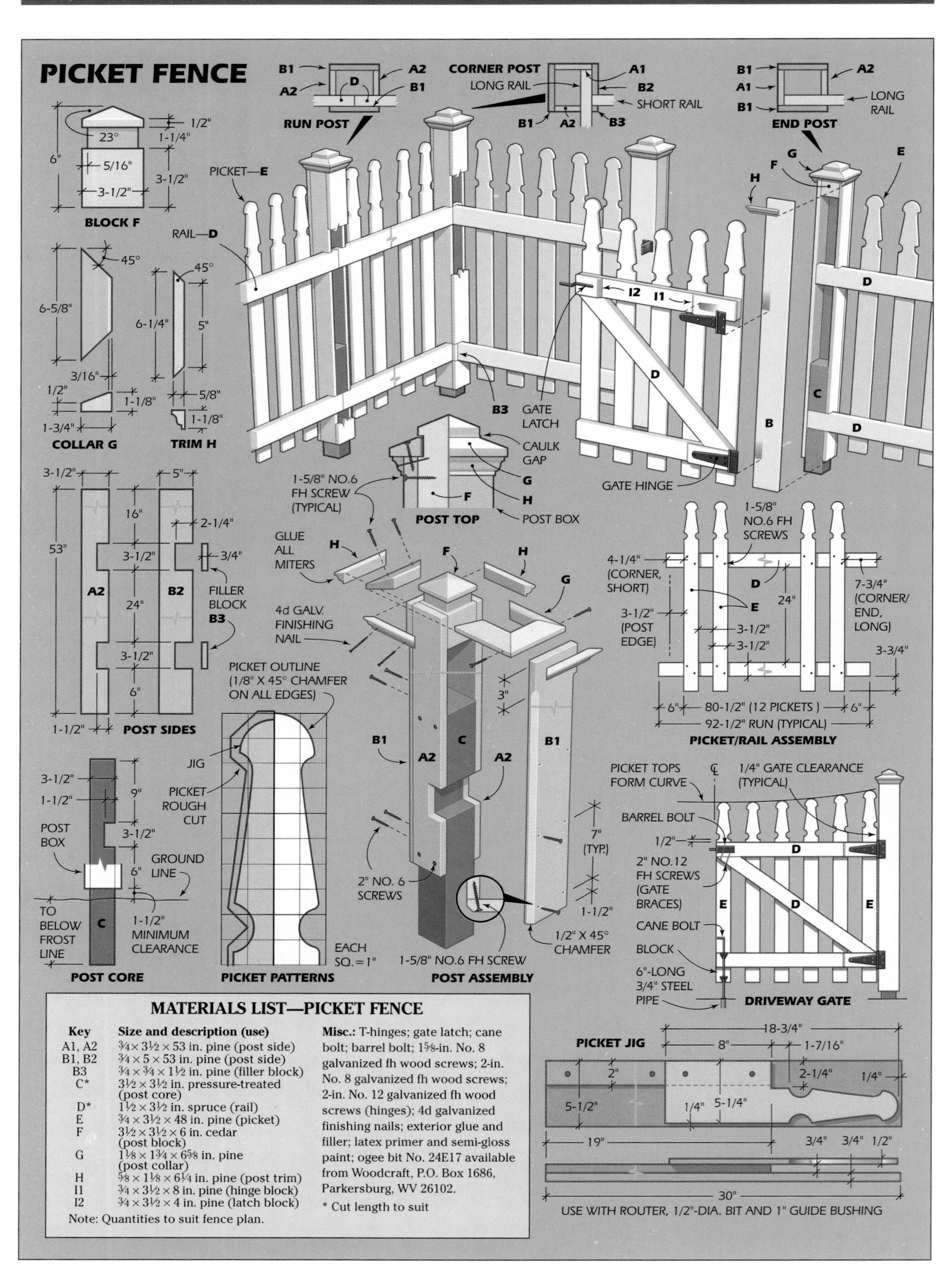

MATERIALS LIST—PICKET FENCE

Key	Size and description (use)
A1, A2	¾ × 3½ × 53 in. pine (post side)
B1, B2	¾ × 5 × 53 in. pine (post side)
B3	¾ × ¾ × 1½ in. pine (filler block)
C*	3½ × 3½ in. pressure-treated (post core)
D*	1½ × 3½ in. spruce (rail)
E	¾ × 3½ × 48 in. pine (picket)
F	3½ × 3½ × 6 in. cedar (post block)
G	1⅛ × 1¾ × 6⅝ in. pine (post collar)
H	⅝ × 1⅛ × 6¼ in. pine (post trim)
I1	¾ × 3½ × 8 in. pine (hinge block)
I2	¾ × 3½ × 4 in. pine (latch block)

Note: Quantities to suit fence plan.

Misc.: T-hinges; gate latch; cane bolt; barrel bolt; 1⅝-in. No. 8 galvanized fh wood screws; 2-in. No. 8 galvanized fh wood screws; 2-in. No. 12 galvanized fh wood screws (hinges); 4d galvanized finishing nails; exterior glue and filler; latex primer and semi-gloss paint; ogee bit No. 24E17 available from Woodcraft, P.O. Box 1686, Parkersburg, WV 26102.

* Cut length to suit

bottoms of the pickets and posts to suit the grade and adjust the bottom rail position accordingly. We preassembled the picket/rail sections in modular lengths of 12 pickets (about 8 ft.) and built shorter, custom sections where necessary. Note that rail lengths change depending on whether the picket/rail assembly is in the middle of a straight run of posts, at a corner or at an end or terminating post.

Carefully lay out the post hole positions and dig the holes below the frost line. Keep the removed soil covered on a tarp or plywood panel until it's time to backfill the holes.

Making the Pickets and Rails

1 Cut stock for pickets, Part E, to size.

2 Make a template of the rough picket profile as shown in the Technical Illustration and use this to mark your work.

3 Gang a number of pickets together, make sure they're aligned and hold them in place with a pipe clamp.

4 Make the 45° cuts with your circular saw *(see Figure A).*

5 Use a sabre saw to make final cuts on the sides of the individual pickets *(see Figure B).*

6 Build the picket jig shown in the illustration to hold each picket while the final shape is routed.

7 Clamp each picket in the jig and use a ½-in. straight bit and 1-in. guide bushing to make the cuts *(see Figure C).*

8 Use a router table and piloted chamfer bit to shape a ⅛-in. bevel around the top and sides of each picket.

9 Cut the rails, Part D, to length based on your plan. Note that the distances from the outer pickets of each section to the rail ends vary depending on the type of post. At an end post the rails extend through the inside of the box. At a corner, the rails of one section run long while the rails of the adjacent section are short and butt against the longer rails. In a continuous run of fence the rails meet at the post centerlines.

10 Preassemble the rail and picket sections, securing each picket with two screws at each rail *(see Figure D).* Use a piece of ply-

FIGURE A: Gang together picket blanks and use a circular saw set at 45° to make most of the cuts for the rough picket shape.

FIGURE B: Finish the rough picket outline by making the longer side cuts on individual pickets with a sabre saw.

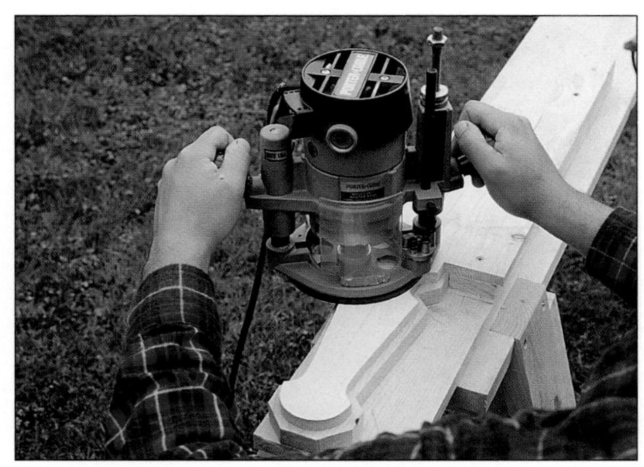

FIGURE C: Use a router, ½-in. bit, 1-in. guide bushing and picket jig to trim the rough pickets to the finished shape.

wood with stops attached to locate the rails. Then use spacer blocks to position the remaining pickets.

Making the Posts

1 Rip 1 × 6 pine to 5 in. wide for the wider post faces, Parts B1 and B2.

2 Cut two 5-in.-wide pieces and two 1 × 4 pieces, Parts A1 and A2, to length for each post.

3 Use the chamfer bit and router table to shape the stopped chamfer on the wider pieces as shown.

4 Cut the rail notches with a sabre saw.

5 Screw two 1 × 4 pieces to a 5-in. piece to make three-sided posts *(see Figure E)*. Bore angled screw pilot holes so the screwheads miss the chamfers. Countersink the holes slightly.

6 Cut the top 4 × 4 blocks, Part F, to length and shape the ends with a miter saw *(see Figure F)*. We used cedar for the top blocks as it was dry and dimensionally stable. If you use pressure-treated stock, you may need to trim the 4 × 4s with a power plane or hand plane so the blocks fit the 3½-in.-sq. box openings.

7 Build a jig to rout the recess around each block as shown.

8 Attach each top block to a post with screws driven through on two adjacent sides *(see Figure G)*.

FIGURE D: Build an assembly table with stops to position rails. Use spacers to locate pickets and attach with two screws per rail.

FIGURE E: Assemble three-sided post boxes. Bore screw pilot holes and drive screws at a slight angle so heads miss the post chamfers.

FIGURE F: Use a power miter box to cut top blocks to size and shape. Set the saw for 23° and trim to centerlines marked on the block end.

FIGURE G: After routing the recess around each block, install three-sided posts and secure with a screw on two adjacent sides.

FIGURE H: Slide three-sided collar around block from open side of post. Secure with screws and attach remaining side with glue and screws.

FIGURE I: Use galvanized finishing nails to secure trim under the post collar around three sides of the box post. Drive two nails per piece.

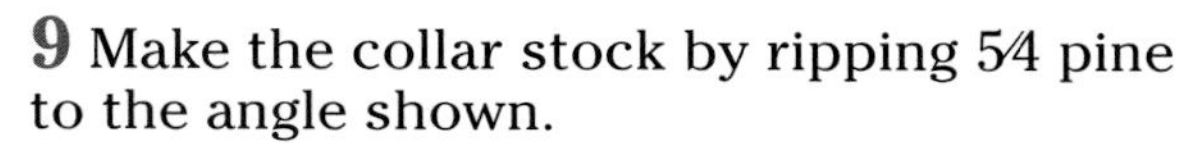

9 Make the collar stock by ripping 5⁄4 pine to the angle shown.

10 Miter each collar piece, Part G, to length.

11 Assemble three-sided collars with screws and exterior glue.

12 Slide each partial collar in place from the open side of the post, align it and secure one side with screws *(see Figure H).*

13 Add the fourth collar piece on each post and screw the collar to the box sides as shown.

14 Use an ogee bit to rout the post cap trim, Part H, in 5⁄4 pine.

15 Miter the trim and install three sides around the box post *(see Figure I).*

16 Cut the 4 × 4 core posts, Part C, to length, check that they will fit in the box cavities and plane the faces if necessary.

17 Place each in its hole and use a line level to locate the notch positions.

18 Cut the notches by making a series of cuts with a circular saw and removing the waste with a chisel *(see Figure J).*

19 Check that the bottom of each box post will end at the correct height above the ground and trim if necessary.

20 Secure the posts to the 4 × 4 cores with four screws on two adjacent sides.

FIGURE J: Cut 4 × 4 post notches by making a series of kerfs with a circular saw. Then remove the waste with a sharp chisel.

Installing and Finishing the Fence

1 After priming the posts, place them in their holes with the open sides facing the inside of the fence.

2 Clamp a picket/rail assembly to each pair of posts.

3 Screw the bottom rail of each section to the notch in the 4 × 4 core post *(see Figure K).*

4 Use braces screwed to stakes to hold the posts upright in the holes. Plumb each post and sight down the fence to make sure they're aligned. When you're satisfied, backfill the holes, tamping down the soil after every few shovels.

FIGURE K: With the posts resting in their holes, install picket/rail assemblies. Screw bottom rails to 4 × 4 posts, then plumb and align fence.

5 Install the remaining post sides with screws driven into the existing post boxes, post cores, top blocks and upper rails.

6 Cover all screwheads with exterior-grade wood filler and nail the remaining piece of trim at the top.

7 Build each gate as a typical fence section, except with end pickets flush with each pair of rail ends.

8 Position the picket tops to create a concave curve as shown and screw them to the rails.

9 Cut a diagonal brace to fit, secure it to the rails with 2-in. No. 12 screws and screw the pickets to the brace.

10 Apply a bead of paintable silicone caulk to the gap around the post collar and along all mitered seams *(see Figure L).*

11 Attach the gates with heavy-duty T hinges, using 2-in. screws where the hinge screw holes are over the fence rails *(see Figure M).*

12 For a double driveway gate, install a cane bolt on one side and attach the second gate to the first with a barrel bolt.

FIGURE L: Apply a paintable silicone caulk in the gap between the collar and top block. Also, caulk any open joints in trim or box sides.

FIGURE M: Clamp support sticks to the gate to help align it with the fence. Use 2-in. screws to secure hinges to fence rails.

13 Fill and sand all remaining holes.

14 We finished our fence with an acrylic primer followed by semi-gloss acrylic white paint.

Text and Photos by Thomas Klenck
Technical Illustration by
Eugene Thompson

Cobblestone Patio and Barbecue

Text and Photos by Merle Henkenius

If you do a lot of outdoor cooking and backyard entertaining, you're probably well aware of your old gas grill's shortcomings—and the limits of your backyard. Chances are, your grill lacks any real counter space, not to mention the versatility to smoke or roast a large turkey. It probably can't generate the searing heat that you need to optimize flavor. And, if you and your grill stand alone by the garage, you've probably realized that an attractive patio might do much to improve your social life.

While there are simple ways to upgrade, like buying a larger portable grill or pouring a concrete patio slab, sometimes it makes sense to step up in style. Instead of a portable grill, we opted for a large, full-featured, built-in gas grill set into a ledgestone enclosure at the end of a cobblestone patio. It's a big project, and pricey, but the results are impressive. And we certainly won't lack for atmosphere or cooking capacity.

Basic Ingredients

We used artificial cobblestone pavers and ledgestones made by Cultured Stone (Cultured Stone Corp., P.O. Box 270, Napa, CA 94559). The 6 x 9-in. pavers come in several thicknesses—we chose the patio paver, which is roughly 2⅜ in. thick. Prices vary, but expect to pay between $6 and $7 per square foot. Our grill's enclosure is made out of concrete block and brick on an 8-in.-deep concrete footing, and the masonry is finished with Cultured Stone's Pro-Fit Ledgestone veneer. We used just over 38 sq. ft. of veneer at about $6.50 per foot. For the countertop, we used five 19 × 20-in. Cultured Stone hearthstones (about $85 total).

Our built-in grill is a Weber-Stephen Summit 475 (Weber-Stephen Products Co., 200 E. Daniels Rd., Palatine, IL 60067). This is a natural-gas unit with 493 sq. in. of primary cooking surface. It delivers up to 50,000 BTUs of heat through four burners and handles slow cooking as well as quick searing for meals large and small. Suggested retail for the Summit 475 is about $1900.

Built-in grills are permanent appliances and are subject to local building codes—especially the gas piping and electrical wiring. Because building codes can vary from region to region across the country, be sure to visit your local code authority to determine material specifications and installation regulations. If you doubt your abilities when it comes to gas or electricity, it's simply a good idea to hire a professional for this part of the job.

Breaking Ground

1 Determine the size and location of your patio and stretch string lines to mark the perimeter. Remove the sod and excavate the area to about 5 in. deep for the gravel and sand bed. If the soil is heavy clay, cut it into small squares and lift it out in chunks *(see Figure A).*

2 To bring electrical conduit and gas piping to the grill area, bore a ⅞-in. hole through the house siding and underlying rim joist.

3 Assuming you have a finished basement, feed a fish tape into the joist space and across the basement ceiling to your service panel.

4 Attach 12-2 w/g cable to the end of the tape and pull it back through the wall.

5 Thread a 4½-in. pipe nipple into the back of a metal exterior box and feed the cable through this nipple *(see Figure B).*

6 Insert the nipple into the hole, screw the box to the wall with 2½-in. deck screws and caulk around the box.

7 Dig a trench between the house and the left side of the future grill enclosure.

8 Most codes require an extension joint—a slip coupling fitted with an O-ring—in the riser between the box and the trench to accommodate any settling. To install the

FIGURE A: After removing the sod, excavate about 5 in. of soil from the patio area. If the soil is heavy clay, cut it into slabs and shovel it out.

FIGURE B: At the house, thread a nipple into the back of the box and feed the cable through this nipple and into the box.

FIGURE C: Run plastic conduit through the trench for the electrical wires. Then glue the riser into the bottom of the extension joint.

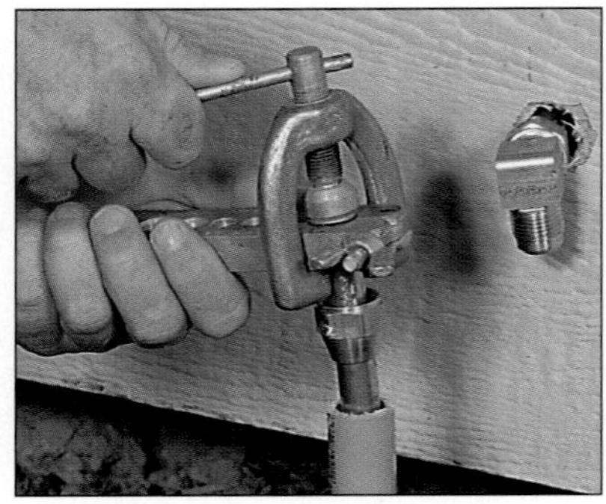

FIGURE D: Use a flaring tool to flare the ends of the copper pipe. Apply pipe dope and tighten the nuts onto a 90° flare fitting.

FIGURE E: Use a rented sand-plate tamper to level and compact the crushed rock. Make at least two passes over the entire bed.

FIGURE F: Lay out a starter row of stones. Then lock them in place with aluminum edging. Drive 10-in. spikes through the edging.

coupling, thread a plastic adapter into the bottom of the box and glue in a short stub of ½-in. plastic conduit. Then glue the top of the coupling to this stub.

9 Run the conduit through the trench, using elbows to make any turns.

10 Glue a 90° elbow to each end to form the risers.

11 At the grill location, simply extend the riser a foot or so above grade and tape it off.

12 On the house side, extend the riser with a length of conduit and glue this length into the bottom of the expansion coupling *(see Figure C)*.

13 We used ⅝-in. soft copper tubing for the gas supply line. Start by boring a ¾-in. hole through the siding and rim joist.

14 Tape the end of the coil and insert the first few feet through the wall. Then, roll out the coil and push it through until you reach the utility room.

15 Cut the pipe with a tubing cutter and slide a ⅝-in. flare nut onto the protruding end.

16 Clamp a flaring tool over the end of the pipe with about ⅛ in. of pipe showing and rotate the handle until the tool bottoms out.

17 Remove the tool and coat the male end of a 90° flare fitting with Teflon pipe tape. Then tighten the nut on the fitting.

18 Lay tubing in a trench between the house and the right-rear corner of the future enclosure, taping off the far end at least 1 ft. above the ground.

FIGURE G: Test fit the cut stones. Then set them in permanently. In this case, we had to notch around the conduit riser.

19 Join the opposite end to the 90° flare fitting *(see Figure D)* and backfill the trench.

20 Caulk around the pipe where it enters the house.

Laying the Patio

1 Fill the patio area with about 3 in. of ¾-in. and smaller gravel and level it with a shovel or rake.

2 Make several passes with a rented tamper *(see Figure E)*.

3 Cover the rock with 1½ to 2 in. of sand and tamp it as well.

4 We chose to lock in the pavers with an aluminum edging made by Brick Stop (Brick

FIGURE H: Frame the enclosure footing and pour the form half-full. Lay three pieces of ½-in. rebar into the concrete and fill the form.

FIGURE I: Use a length of straight 2 × 4 to screed off the concrete in a sawing motion. Then use an edging tool to dress the perimeter.

Stop Corp., 363 Canarctic Dr., Toronto, Ontario, Canada M3J-2P9). Using a straight line of reference—in our case the concrete steps of the house—set one paver against another. When you've reached the width you want, set the aluminum edging against the edge of the stones and nail it down with spikes—at least one every 12 in. *(see Figure F)*. When working around a corner, simply bend the edging and continue. When you have a 3-ft. area of stones, work from those stones to avoid any traffic damage to the sand bed.

5 If you reach a point where a stone needs to be cut, measure it carefully and shape it with a circular saw fitted with a masonry blade. Test fit the cut piece and set it in place *(see Figure G)*.

6 After you have all but the row of stones adjoining the grill enclosure in place, you're ready to build the enclosure's footing. But before you do, sweep a little sand over the stones to help stabilize them.

Building the Grill Enclosure

1 Build a 2 × 10 frame with an inside dimension of 66 × 36 in.

2 Remove the soil about 8 in. overall and about 2 in. deeper at the perimeter so the form edge is at the correct height above grade. Space the form 9½ in. from your last row of pavers to leave room for the final row.

3 Use a 4-ft. level to check that the top of the form is level and flush with the top of the paving stones.

4 Screw stakes to the form, checking across and corner to corner with the level before fastening each one.

5 We ordered a strong half-yard of 2500 psi concrete and carried it to the form in a wheelbarrow. Pour in about half the concrete, then lay in three lengths of ½-in. rebar *(see Figure H)* and fill the form.

6 Screed off the excess with a 2 × 4 *(see Figure I)* and run an edging tool around the perimeter.

7 After two days, strip the forms, fill the space next to the patio with crushed rock and sand and lay the last row of stones.

8 Install aluminum edging along the finished edge next to the footing *(see Figure J)*.

9 The grill mounts on a stainless steel insert sleeve. To suit the sleeve and provide adequate counter space we planned an enclosure that was four concrete blocks wide (64 in.) by two blocks deep (32 in.). We actually needed 33 in. of depth to accommodate the sleeve, so we stretched the side with wider mortar joints. Along the patio, set the blocks 1½ in. inward from the edge to prevent the ledgestone veneer from overhanging the pavers.

10 Mix a batch of mortar and set the front row of blocks. Use a 4-ft. level to keep the blocks straight and level *(see Figure K)*.

11 Lay the rest of the first course, leaving a half-block opening at the right-rear corner for the gas connection and another to vent the enclosure.

12 Continue up two courses, leaving two front blocks out on the second course for the stainless steel sleeve.

13 To support the sleeve, set two blocks inside the enclosure, just behind the front opening *(see Figure L).*

14 Lay 4-in. cap blocks on the two front blocks and two interior blocks, and level them *(see Figure M).*

15 To achieve the correct total height—12 in. under the bottom of the sleeve and 21⅛ in. above it—you'll need three courses of standard block, one course of concrete brick and one course of 4-in. cap block.

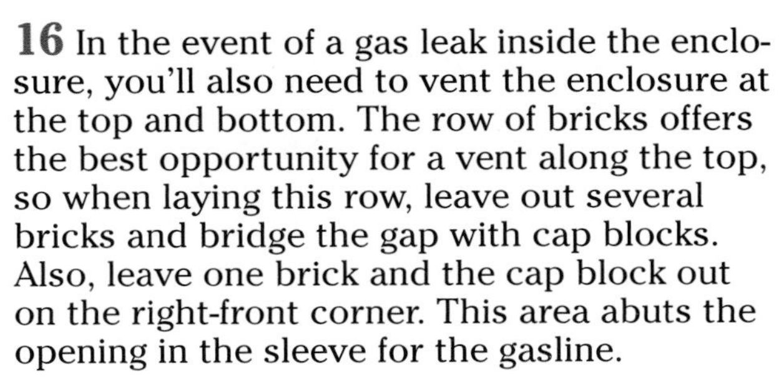

16 In the event of a gas leak inside the enclosure, you'll also need to vent the enclosure at the top and bottom. The row of bricks offers the best opportunity for a vent along the top, so when laying this row, leave out several bricks and bridge the gap with cap blocks. Also, leave one brick and the cap block out on the right-front corner. This area abuts the opening in the sleeve for the gasline.

17 Assemble the insert sleeve panels and fit the grill-support bars into the holes in the sleeve sides *(see Figure N).*

18 Set the assembled sleeve into the enclosure and check the clearance for the gasline.

19 Lay a partial brick in this area and then lay the final cap block.

20 To provide counter-top support, lay up a single-width column of bricks next to the sleeve and centered in the opening.

FIGURE J: Finish laying the cobblestone pavers between the footing and patio. Lock them in place with aluminum edging.

FIGURE K: Lay the first course of enclosure concrete blocks on the footing. Use a 4-ft. level to straighten and level the blocks.

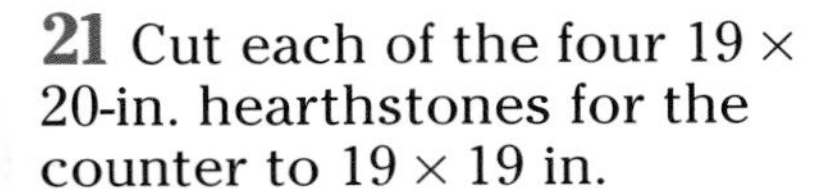

21 Cut each of the four 19 × 20-in. hearthstones for the counter to 19 × 19 in.

22 Cut a ⅜ × 3-in. notch in the front two hearthstones so they fit around the edge of the sleeve.

23 Prepare two pieces 10¾ × 15½ in. for the back of the grill.

24 Lay a loose bed of mortar on the cap blocks *(see Figure O),* tip the hearthstones under the rim of the sleeve and settle them in place, tapping them down until they're level.

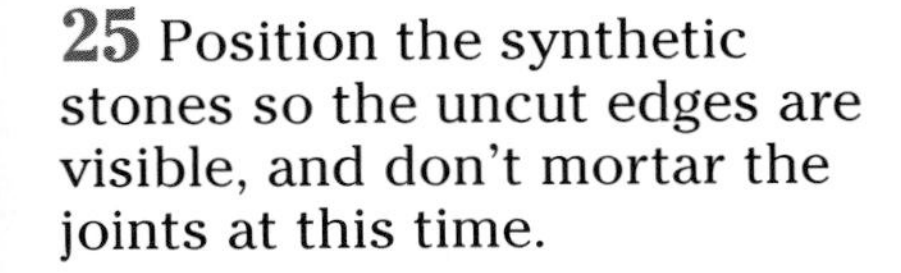

25 Position the synthetic stones so the uncut edges are visible, and don't mortar the joints at this time.

26 To install the ledgestone veneer, first apply at least ½ in. of mortar to one of the front corners with a trowel.

27 Press the first corner piece into the mortar. Do the same with the course above this one, alternating the long end of the piece *(see Figure P).*

FIGURE L: Leave an opening in the second course and set two interior blocks to support the sides of the stainless steel grill sleeve.

FIGURE M: Finish the sleeve supports with 4-in. cap blocks. Level in all directions to make sure the grill sits level in its enclosure.

FIGURE N: Screw together the stainless steel panels of the support sleeve. Lock in the two grill-support bars before fastening the back.

FIGURE O: Lay a loose bed of mortar on top of the cap blocks to accept the hearthstones that make the enclosure's countertop.

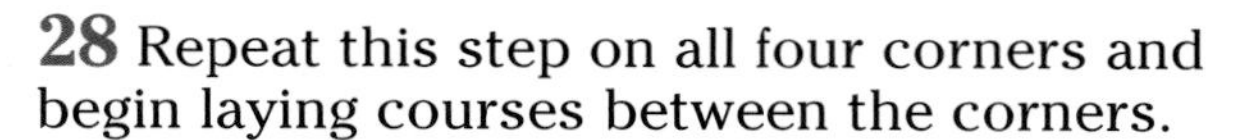

28 Repeat this step on all four corners and begin laying courses between the corners.

29 Where the stones meet, cut a piece to fit, mortar the back of it *(see Figure Q)* and press it in place.

30 Continue until you've laid the ledgestone up to the bottom of the hearthstones. Leave the veneer off a small area above the gasline access opening for the gas regulator that comes with the grill.

31 Finally, install vent panels over the enclosure's vent openings *(see Figure R)*. We used 4-in. soffit vents, because they're the same depth as the veneer. Anchor or caulk them in place.

Installing the Grill

1 Have a helper on hand to ease the grill into the steel support sleeve.

2 Install the door that covers the storage area below the grill *(see Figure S)* and add the control knobs and other accessories.

3 Tape off the leading end of the grill's flexible gas tube and feed it through the opening in the right side of the support sleeve.

4 Concrete is abrasive, so slide a length of foam-rubber pipe insulation over the tube inside the masonry enclosure.

5 Attach the upper end of the tube to the burner manifold.

6 Use plastic anchors to mount the gas regulator to the exposed block just above the gasline opening.

7 Cut the copper pipe to length, install a flare fitting and gas valve, and pipe the line into the right side of the regulator with black-iron pipe.

8 Using a light coating of pipe joint compound, connect the grill's tube to the brass regulator fitting *(see Figure T)*.

9 The connection indoors will depend on the type of gas system you have. If it's low-pressure gas—2 to 4 ounces—with the regulator on the gas meter, just connect the copper to the black-iron piping near your furnace. If it's a medium-high pressure system—about 2 pounds—you'll find the regulator indoors, usually near the furnace. All appliances need to be connected to the system after this regulator.

10 Shut the gas off at the regulator.

11 Loosen the union and thread the nipples from the low-pressure side of the regulator.

12 Install a short nipple, a 90° L and a T and connect this assembly to the rest of the yoke with the union. Because we are splicing in a larger feed line to the grill, our gas code requires the new T feeding this line be oversize as well. To accomplish this, we used a ¾-in. T, fitted with two ½-in. threaded bushings, then threaded our ½-in. black nipples into these bushings.

13 From the top of the T, install a short nipple, a gas valve and a flare fitting.

14 Flare the copper feed line and tighten the flare nut onto the fitting.

FIGURE P: Press two corner ledgestone pieces into the wet mortar, making sure to alternate the long ends of the pieces.

FIGURE Q: Lay the ledgestone from the corners to the center. Cut the last piece to fit, apply mortar and firmly press it into the opening.

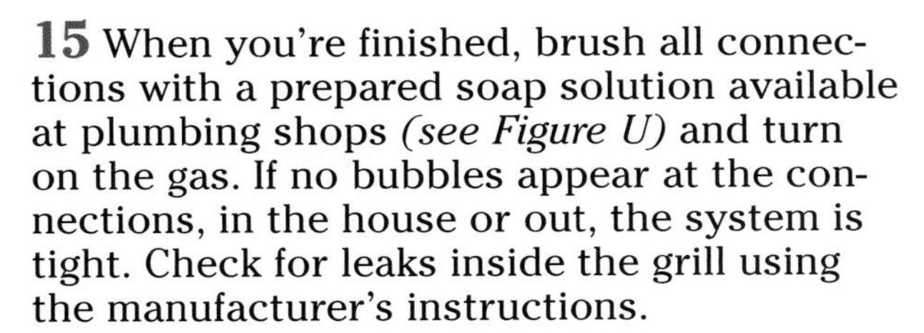

15 When you're finished, brush all connections with a prepared soap solution available at plumbing shops *(see Figure U)* and turn on the gas. If no bubbles appear at the connections, in the house or out, the system is tight. Check for leaks inside the grill using the manufacturer's instructions.

16 To complete the electrical supply, first glue an exterior box to the conduit riser and screw the back of the box to the enclosure using an anchor kit *(see Figure V)*.

17 Pull black, white and green insulated wires through the underground conduit and tie all like-colored wires together in the box at the house using twist connectors.

18 Ground this box and seal it with a weather-tight coverplate.

19 Install a ground fault circuit interrupter (GFCI) receptacle and coverplate at the grill.

20 At the service panel, shut the power off at the main disconnect and remove the cover.

21 Attach the white circuit wire to the neutral bus bar and the ground wire to the grounding bus.

22 Snap a 20-amp breaker into the panel and tighten the black wire under its terminal screw.

FIGURE R: Vent openings are required in case of a gas leak. To keep pests from entering the enclosure, cover the openings with vent panels.

FIGURE S: After setting the grill in place on the insert, install the door that covers the storage area under the grill.

FIGURE T: Connect the grill's gas tube to the regulator behind the grill. Protect the copper riser with a length of plastic conduit.

FIGURE U: Join the new gasline to the existing piping near the furnace. Test each connection with a code-approved soap solution.

Finishing the Stonework

1 Begin grouting the joints between the hearthstones by taping off both sides of each joint, to keep from spilling mortar on the stones.

2 Using fairly wet mortar, fill each joint and tool it with a tuckpoint trowel *(see Figure W)* or a joint striker.

3 After a two-day wait, use silicone to caulk the entire perimeter of the sleeve *(see Figure X).*

4 After the caulk has cured, brush on two thin coats of clear concrete sealer, waiting 2 hours between coats.

FIGURE V: After the mortar has cured, mount the new electrical box to the veneer of the enclosure with a plastic anchor kit.

FIGURE W: Grout the hearthstone with mortar and tool it with a striker or tuckpoint trowel. Tape off the stones to avoid mortar stains.

FIGURE X: Apply clear silicone caulk around the entire perimeter of the sleeve. When it cures apply two coats of sealer to the counter.

PHOTO BY DES AND JEN BARTLETT/NGS/IMAGE COLLECTION

Double Decker Porch

For those who live in clement climes, winter can be quite a relief, a real respite from high temps and air-conditioned isolation. But for the rest of us the season brings blessings that are more mixed. For outdoor enthusiasts, there are any number of activities that can take the edge off the long winter months. But when it comes right down to it, the driveway still has to be shoveled, the car still has to be treated with more care than many of us would like, and the lure of Sunday football games broadcast from sunny locales all serve to remind us that spring really is a better idea. It's a one-way ticket outside, where the air is fresh, the sun is warm and it seems like you can accomplish just about anything before booking return passage.

PHOTO BY THOMAS KLENCK

If you're looking for a way to get outside, stay out of trouble and improve your home, all at the same time, then this massive double decker porch project may be for you. So, get out your tools and some of your ambition, and have at it. And don't forget, after a productive summer, the cold air, your easy chair and a good game are right around the corner.

Make no mistake about it: This is a big project. But if you approach it methodically you should be able to wrap it up before the snow flies with enough time left over to watch the leaves change from your porch rocker.

This elevated porch was designed for a house with a full-height walk-out basement. This yielded a traditional covered porch on top and a covered play area for the kids below. But the same construction techniques apply if you want to build the porch at grade level. All you have to do is shorten the posts.

Making the Porch Foundation

1 Lay out the location for the four foundation piers (see deck plan view in the Technical Illustration).

2 Excavate the holes below the frost line to accommodate your foundation piers.

3 For this job we used a plastic footing form called a Bigfoot (F&S Manufacturing, RR#1 Chester Basin, Nova Scotia, Canada B0J 1K0) that has a 28-in.-dia. base. This is a great product because it allows you to pour the footing and pier column at the same time. To use it, just screw a cardboard pier form to the top of the plastic funnel *(see Figure A)*.

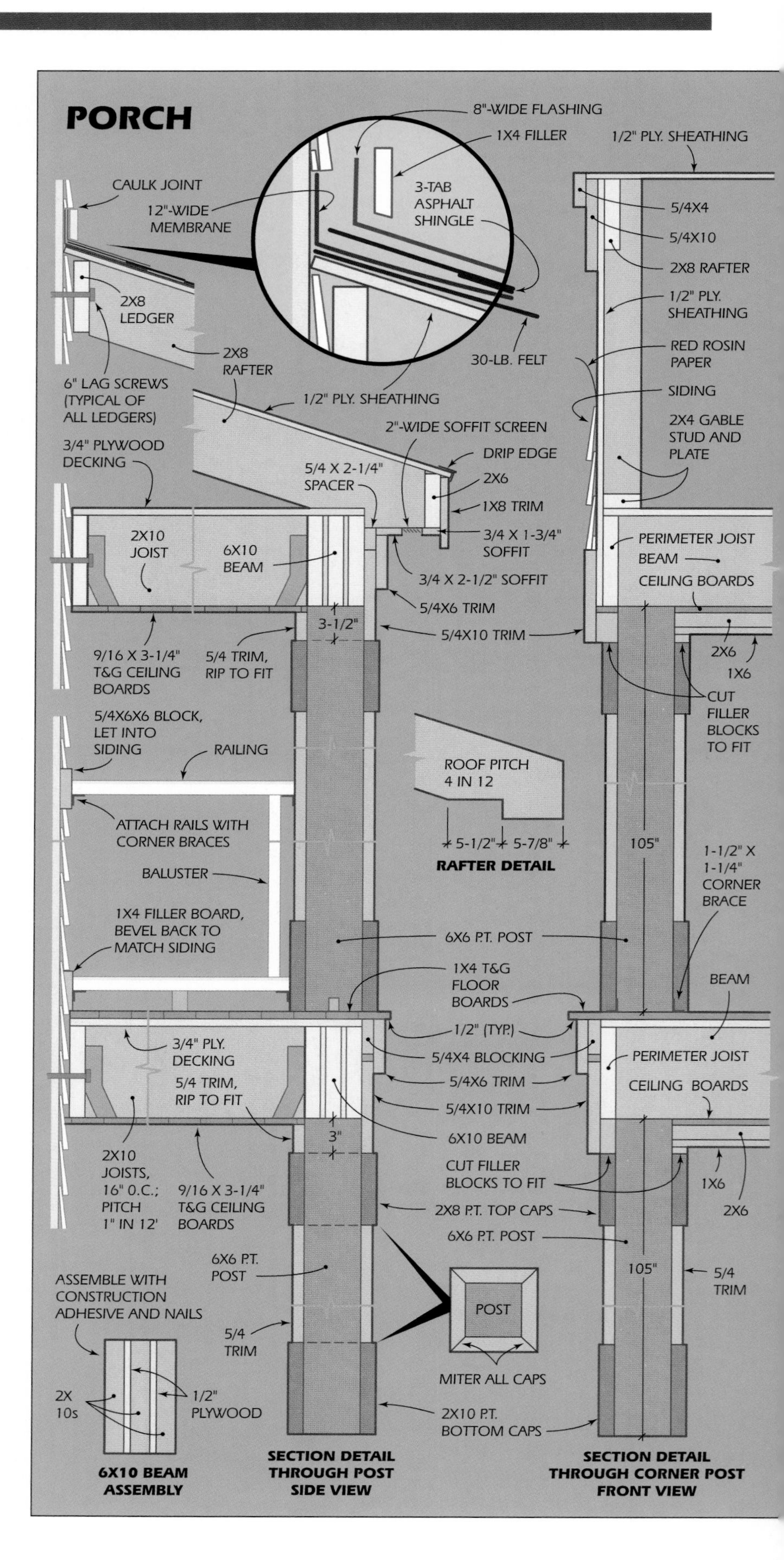

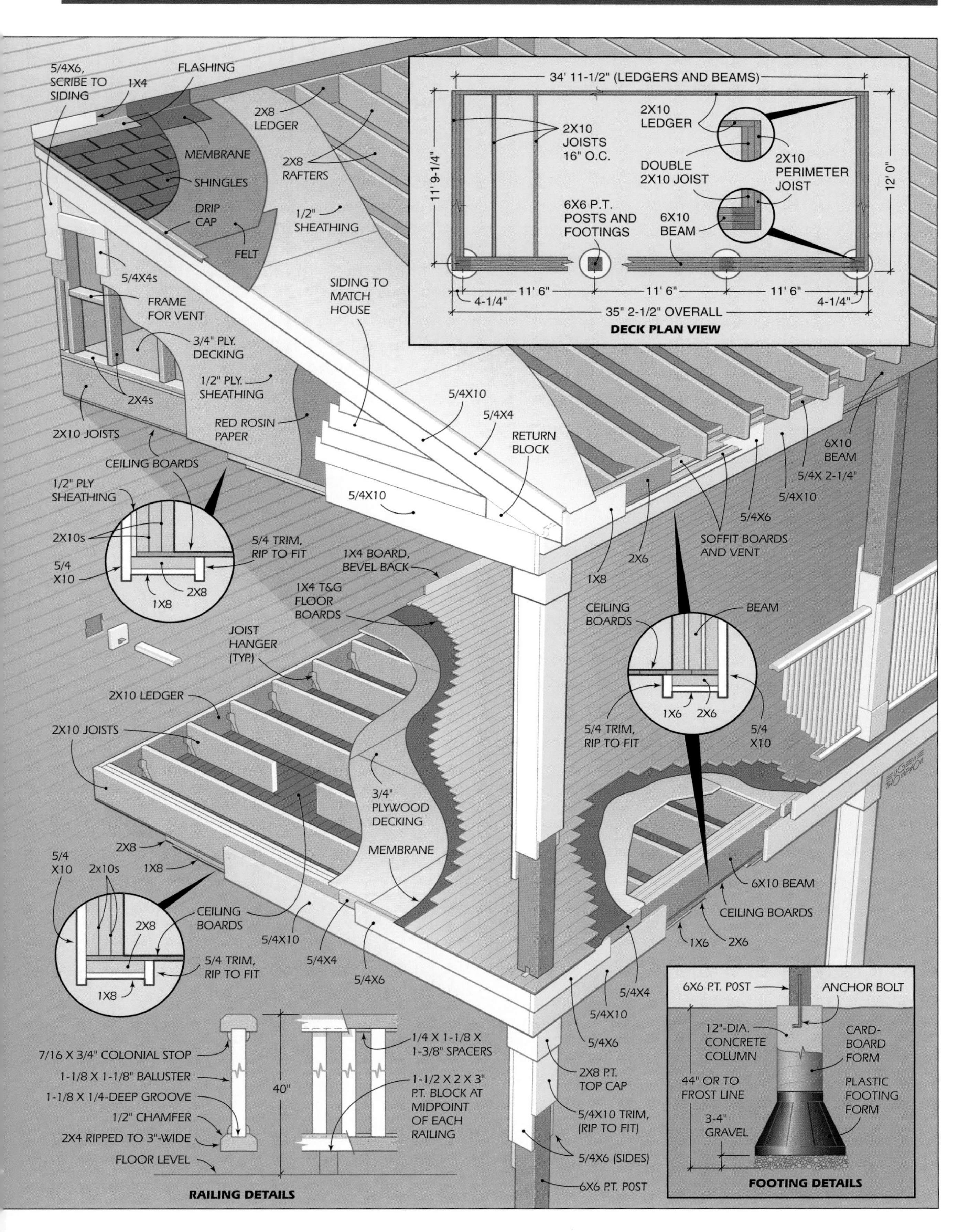

5/4X6, SCRIBE TO SIDING
1X4
FLASHING
2X8 LEDGER
MEMBRANE
SHINGLES
2X8 RAFTERS
DRIP CAP
1/2" SHEATHING
FELT
5/4X4s
FRAME FOR VENT
SIDING TO MATCH HOUSE
3/4" PLY. DECKING
1/2" PLY. SHEATHING
2X4s
RED ROSIN PAPER
2X10 JOISTS
CEILING BOARDS
5/4X10
5/4X4
RETURN BLOCK
6X10 BEAM
5/4X 2-1/4"
5/4X10
5/4X6
1/2" PLY SHEATHING
2X10s
5/4 TRIM, RIP TO FIT
5/4 X10
2X8
1X8
1X4 BOARD, BEVEL BACK
1X4 T&G FLOOR BOARDS
SOFFIT BOARDS AND VENT
2X6
1X8
CEILING BOARDS
BEAM
JOIST HANGER (TYP.)
2X10 LEDGER
2X10 JOISTS
1X6
2X6
5/4 TRIM, RIP TO FIT
5/4 X10
3/4" PLYWOOD DECKING
MEMBRANE
6X10 BEAM
CEILING BOARDS
2X8
5/4 X10
2x10s
1X8
CEILING BOARDS
2X8
5/4X10
5/4X4
5/4X6
5/4 TRIM, RIP TO FIT
1X8
1X6
2X6
5/4X4
5/4X10
5/4X6
34' 11-1/2" (LEDGERS AND BEAMS)
2X10 LEDGER
2X10 JOISTS 16" O.C.
DOUBLE 2X10 JOIST
2X10 PERIMETER JOIST
11' 9-1/4"
12' 0"
6X6 P.T. POSTS AND FOOTINGS
6X10 BEAM
11' 6"
11' 6"
11' 6"
4-1/4"
4-1/4"
35" 2-1/2" OVERALL
DECK PLAN VIEW
7/16 X 3/4" COLONIAL STOP
1-1/8 X 1-1/8" BALUSTER
1-1/8 X 1/4-DEEP GROOVE
1/2" CHAMFER
2X4 RIPPED TO 3"-WIDE
FLOOR LEVEL
40"
1/4 X 1-1/8 X 1-3/8" SPACERS
1-1/2 X 2 X 3" P.T. BLOCK AT MIDPOINT OF EACH RAILING
RAILING DETAILS
2X8 P.T. TOP CAP
5/4X10 TRIM, (RIP TO FIT)
5/4X6 (SIDES)
6X6 P.T. POST
6X6 P.T. POST
ANCHOR BOLT
12"-DIA. CONCRETE COLUMN
CARD-BOARD FORM
44" OR TO FROST LINE
PLASTIC FOOTING FORM
3-4" GRAVEL
FOOTING DETAILS

4 Drop the assembled form in the hole and plumb it in place, then backfill around it.

5 Pour the concrete. When the concrete is still soft, put an anchor bolt in the top of each pier to receive its post.

Building the First Floor

1 Cut the 6 × 6 posts to length, remembering to cut each 1 in. short so the floor will be pitched away from the house for water runoff.

2 Lay out the point where the anchor bolt falls in each post and bore a bolt clearance hole.

3 Lift each post in place *(see Figure B).*

4 Nail braces near the top of two adjacent post sides, then drive stakes into the ground to receive these braces.

5 Plumb each post from both directions and nail the braces.

6 Cut the house ledger board to length and lay out the location of all the floor joists.

7 Nail joist hangers at each layout mark, and bore bolt clearance holes in the ledger to match the locations of all the house studs or the house rim joist. Stagger these holes and make sure you have at least one every 16 in. along the length of the board.

8 When the holes are finished, lift the ledger into place and attach it with the lagbolts and washers *(see Figure C).*

9 Lay out one inside beam member of the first floor beam to match the joist locations on the house ledger and nail the joist hangers in place.

FIGURE A: Attach a concrete form tube to the plastic footing form with four drywall screws. Space screws evenly around the tube.

FIGURE B: Mark the location of the anchor bolt on each post bottom. Then bore a clearance hole and lift the post onto the bolt.

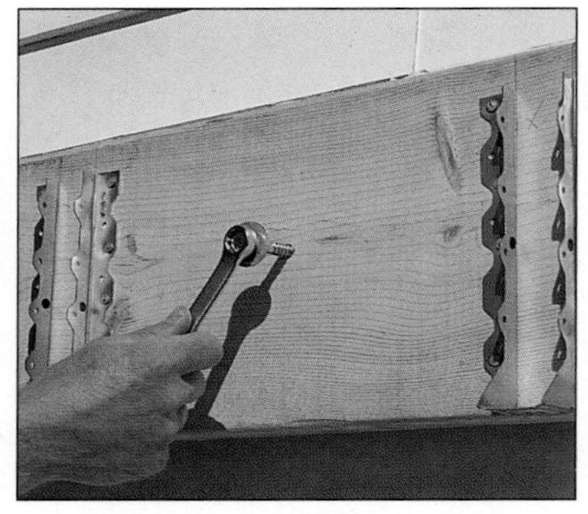

FIGURE C: Lay out the ledger board for the first floor deck and nail on the joist hangers. Attach the ledger to the house with lagbolts.

FIGURE D: Lay out and attach joist hangers to the inside member of the first floor beam. Lift this board onto the posts.

FIGURE E: Cut the first floor joists to length and slide them in place between the joist hangers on the beam and the house ledger.

FIGURE F: Nail plywood over the first floor joists and cover it with roofing membrane. Then nail the floorboards in place.

FIGURE G: Paint the floorboards and lay out the second floor post positions. Cut the posts to length and toenail each to the floor.

FIGURE H: Lay out the rafter ridgeboard and attach it to the house with lagbolts. Make sure the bolts hit the wall studs.

FIGURE I: Lay out and cut a pattern rafter and check its fit. Then use it to carefully mark both ends of all the other rafters.

FIGURE J: Nail the rafters at the bottom first. Make sure the birdsmouth fits tightly before toe-nailing the rafter to the deck.

10 Lift the beam member onto the posts *(see Figure D).* Toenail this board into the posts and then add the other two inside members in the same way.

11 When all are nailed in place, put construction adhesive on the outside surface. Then nail plywood spacers as shown in the drawing to these beam members.

12 Complete the beam by gluing and nailing together the other beam members.

13 Cut all the floor joists to length and lift them into the joist hangers on the ledger and the beam *(see Figure E).* Be sure to fill all the joist hanger holes on both sides and ends of each joist with properly sized joist hanger nails.

14 Begin laying the floor plywood by measuring 48 in. out from the house on both ends on the deck. Snap a chalkline between these two points.

15 Apply construction adhesive to the top edges of the joists between the house and the chalkline.

16 Carefully place the tongue-and-groove plywood onto the joists with the good side up and the groove edge pointing out.

17 Nail the first sheet every 6 to 8 in. along each joist.

18 Continue laying and nailing the rest of the plywood in the first course.

19 When you're done, apply construction adhesive to the top of the joists for the next course of plywood.

20 Carefully lay the first sheet in the second row against the first sheet in the first row so the tongue and groove match up. Then tap the two sheets together using a sledgehammer and a scrap 2 × 4 along the grooved edge of the second sheet. The 2 × 4 protects the groove from being damaged by the sledge.

21 Continue in the same manner until all the plywood is cut, driven and nailed in place.

22 Install the trim boards around the first floor deck and the first floor posts.

23 We used roofing membrane between the plywood and the finished floorboards. Just install the membrane according to the package directions.

24 Back prime the floorboards before nailing them in place.

25 Cover the membrane with floorboards, nailing through the tongue edge and into the joists below *(see Figure F).*

26 Apply a coat of self-priming deck enamel to the floorboards.

Building the Second Floor and Roof

1 Once you have a coat of enamel on the floor, lay out the post positions.

2 Cut the posts to length and toenail them into place *(see Figure G).*

Building the ceiling is nothing more than a repeat of the first floor deck.

3 Once you have a platform to work on, lay out the roof ridgeboard and bolt it to the house *(see Figure H).*

4 Use a framing square to lay out a rafter pattern.

5 Cut the pattern and then try it in several places along the ridge and the beam to make sure both ends fit well.

6 When satisfied with the pattern, use it to trace the cut lines on all the other rafters *(see Figure I).*

7 Cut the rafters to shape and then lift each in place and nail at the birdsmouth end first *(see Figure J).*

8 Toenail the other end into the ridgeboard *(see Figure K).*

9 Nail the roof sheathing over the rafters.

10 Cover the roof sheathing with asphalt felt. This will protect the framing from rain damage.

FIGURE K: Once the rafter bottom is nailed, toenail the top in place. Use one nail at the ridge and two at each side.

FIGURE L: Nail a bottom plate to the deck at both gable ends. Cut the gable studs to length and toenail them in place.

FIGURE M: Cover the gable studs with sheathing and nail the bottom trim board in place. Use 12d galvanized finish nails.

FIGURE N: Cut the vertical trim board to length and hold it against the house. Scribe the siding to the board and cut it with a sabre saw.

11 Trim the rafter tails and the second floor posts as shown in the Technical Illustration.

12 Now, you need to fill in the gable end framing. Start by nailing a bottom plate along both gable ends.

13 Lay out and cut the gable studs to size.

14 Toenail them into the bottom plate *(see Figure L)* and through the sides of the gable rafters.

15 Cover these studs with plywood sheathing.

16 Add the trim boards to both gable ends as shown in the illustration. Start with the horizontal board along the bottom of each gable *(see Figure M)*.

17 Add the rake boards and finish the trim by scribing in place the vertical board next to the house *(see Figure N)*.

18 Cover the exposed sheathing with building paper and nail on the siding *(see Figure O)*.

19 Install aluminum drip edge around the entire roof perimeter.

20 Add a course of roofing membrane along the eave edge. Roll the membrane out *(see Figure P)*, remove the backing paper and smooth the membrane onto the felt.

21 Begin installing the shingles by first nailing a starter strip along the eave edge *(see Figure Q)*. This strip is nothing more than a row of standard shingles turned upside down.

22 Install the rest of the shingles, staggering the joints by half a tab as you work up the roof.

23 Once you get to the top of the roof, you'll have to install flashing between the house and the roof to make the joint watertight. We removed some house siding, and then installed a layer of membrane, followed by a row of shingles and a layer of aluminum flashing. The flashing and membrane were protected against the house with a trim board that was nailed in place over the flashing layers. All the nailheads were set and covered with silicone caulk, as was the top edge of the trim where it fit under the siding.

FIGURE O: Cover the gable sheathing with building paper. Then nail the siding boards in place using galvanized siding nails.

FIGURE P: Cover the roof plywood with 15-lb. asphalt felt. Then add a layer of roofing membrane to the eave edge of the roof.

FIGURE Q: Begin the roof shingles by nailing a starter course in place. Nail these shingles upside down along the eave.

FIGURE R: Cut the railing stock to width and length. Then rout a groove in all rails to accept the ends of the balusters.

FIGURE S: Cut the balusters and spacing blocks to size and nail them into the rail grooves using 6d galvanized finish nails.

FIGURE T: Cut the rough opening for the door through the house wall. Then tip the prehung door assembly into place.

Completing the Porch

1 While the porch railings can be fabricated in place, it's much easier to build them on sawhorses and then install them as assembled units. Begin by cutting the rails to length and width.

2 Rout a groove in each rail to accept the balusters *(see Figure R).*

3 Cut all the balusters and spacer blocks to size and nail them into the rail grooves *(see Figure S).*

4 Spread silicone caulk along the sides of the spacer blocks and nail the colonial stop in place.

5 When the caulk is dry, prime all the parts and lift the assemblies in place between the porch posts.

6 Attach them to the posts at both ends of each rail with galvanized corner braces.

7 We used beaded tongue-and-groove fir boards for both ceilings on this porch. We simply nailed them to the bottom edge of the joists using 6d finish nails driven through the tongue.

8 Create your access door opening into the house. Because we replaced a wide window with the doorway, our job was simplified. The existing header opening was wide enough for the door so we didn't have to replace the header. We simply cut the window opening to the floor and furred out the existing framing a bit.

9 Install the door in the opening *(see Figure T).* We used a prehung Ever-Strait Classic steel door, model No. E-238, and a Glendale lever lockset from Pease Industries (7100 Dixie Hwy., Fairfield, OH 45014). Be sure to check for square and plumb before nailing the flanges to the framing.

10 Trim out the outside and inside of the door to match your existing trim details.

11 Finish up by giving all the primed surfaces two topcoats of high-quality exterior paint.

Text and Photos by Steven Willson
Technical Illustration by
Eugene Thompson